PHYSICS NOTES
Number 1

PHYSICS NOTES

EDITED BY

Philip W. Anderson, Arthur S. Wightman, and Sam B. Treiman

1. RENORMALIZATION GROUP
by Giuseppe Benfatto and Giovanni Gallavotti
(1995)

RENORMALIZATION GROUP

Giuseppe Benfatto and Giovanni Gallavotti

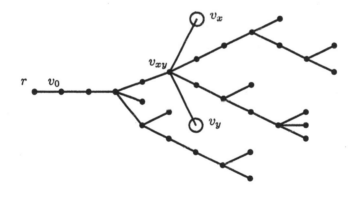

Physics Notes

Princeton University Press
Princeton, New Jersey

Library of Congress Cataloging-in-Publication Data

Benfatto, Giuseppe, 1944–
Renormalization group / Giuseppe Benfatto and Giovanni Gallavotti.
p. cm. — (Princeton physics notes)
Includes bibliographical references and indexes.
ISBN 0-691-04447-3 (alk. paper). — ISBN 0-691-04446-5 (pbk. : alk. paper)
1. Renormalization group. 2. Critical phenomena (Physics)
I. Gallavotti, Giovanni. II. Title. III. Series.
QC20.7.R43B46 1995
530.1'33—dc20 95-10533

The publisher would like to acknowledge the authors of this volume for providing
the camera-ready copy from which this book was printed

Princeton University Press books are printed on acid-free paper and meet the
guidelines for permanence and durability of the Committee
on Production Guidelines for Book Longevity
of the Council on Library Resources

Printed in the United States of America
by Princeton Academic Press

10 9 8 7 6 5 4 3 2 1

10 9 8 7 6 5 4 3 2 1 (Pbk.)

Contents

Contents

Preface

One of us (GG) gave a series of six lectures (two hours each) at the IIIe cycle de la Suisse Romande in January/February 1990. The notes on the lectures gave us the idea of writing the present book. Our intent has been to stress the ideas and the conceptual developments, referring to the literature for the proofs. Even so we realize that we were forced to make several compromises with the formalism (particularly in Chapter 10, obviously the heart of the matter), and, as a consequence, this book assumes that the reader has some familiarity with classical quantum field theory. What is really required is that the reader has worked out the details of some classical perturbation theory result (e.g., the Compton scattering in QFT, the removal of some divergences in one-loop radiative corrections, and some elementary calculations in the field theory approach to many-body theory).

Therefore the book is dedicated to scholars wishing to reflect on some details of the foundations of the modern renormalization group approach and to think of them in the light of the few known rigorous results. We use them here to provide support for an approach that has proven useful in many problems.

We do not take the point of view of mathematical physics. Rather we expose how the renormalization group looks to us as physicists: this means the achievement of a coherent perturbation theory based on second order (or lowest-order) calculations. The contribution of mathematical physics to the subject has been, we believe, the clarification of the foundations (certainly clear to the initiators of the subject, but less so to many others) and to provide proofs — often only partial — that the higher-order corrections do not matter.

Therefore we do not dwell at all on the successes of the renormalization group methods as a mathematical physics subject (hierarchical model, φ_3^4 scalar field theory, $d = 2$ Coulomb gas and dipole gas, $d = 4$ critical point, to mention a few) or on the general $n!$ bounds (such as (8.28) below). This is so not only because it would expand our work almost without limits, but also because it is time that such results became accepted (without the need to explain them from the beginning to *audiences who do not really care), allowing us a discussion of the renormalization group by keeping the rigorous results in the background, in a context that illustrates the key ideas without requiring sophisticated (or perhaps just complicated) mathematical tools and technical analysis. We use the mathematical results only by occasionally quoting them to say that the higher order corrections do not change the picture.*

The material collected here is fiarly standard except for the part on the Bose condensation, which we added because it is a summa of all the previously illustrated ideas and in order to stimulate a discussion on one of the most controversial subjects in condensed matter theory.

We restrict our attention to some typical problems that can be formulated as field theories, but the methods are quite general. The ideas reflect the approach that was developed in Roma during the last 15 years (eighteen as of this writing). We are indebted to many for encouraging us by simply paying attention to our efforts and for often providing suggestions that we have incorporated, possibly (but not intentionally) without reference, because ours was, at least at the beginning, a fascinating joint cultural development. While this work shows us how much lower the results have been compared to the expectations, we find it useful to report our point of view to a wider audience.

The aim of the lectures in Lausanne was to present the results of the renormalization group concisely and in a nontechnical form. The hope was to convey the enthusiasm evoked in the lecturer when he realized — finally getting a flash of ideas from the papers of [DJ], [D], [GJ], [BS1], [WF] whose deep significance had remained obscure to him for long years — that mean field theory of phase transitions had finally been surpassed and not only nontrivial critical exponents were naturally emerging, but also at the same time, and for the same reason, renormalization theory of relativistic QFT became fairly transparent. The technicalities of the exposition that we could not avoid both in the original version and, more so, in the present probably obscure the original intention, but we hope that some traces of the revelation still survive.

We are greatly indebted to Professor Arthur Wightman for his interest, his comments, and his encouragement as well as for making possible the publication of this book through Princeton University Press.

Roma, 1995

RENORMALIZATION GROUP

Chapter 1
Introduction

The notion of renormalization group is not well defined. It arises in theories in which a prominent role is played by scale invariance or covariance properties, of various quantities, with respect to a noninvertible transformation of coordinates. The noninvertibility is an essential feature; it is supposed to allow one to reduce effectively the difficulty of the problem.

Examples of problems that have been treated implicitly or explicitly via such techniques are the following:

1. The KAM theory of Hamiltonian stability
2. The constructive theory of euclidean fields
3. The universality theory of the critical point in statistical mechanics
4. The onset of chaotic motions in dynamical systems
5. The convergence of Fourier series on the circle
6. The theory of the Fermi surface in Fermi liquids

To the above one can add a less-established problem, which we present here because it seems to us that it summarizes and merges many of the ideas illustrated in this book:

7. The Bose condensation.

The term *renormalization group* started to be used in the above sense with the work of Di Castro and Jona-Lasinio [DJ], and the method was developed to its major successes in the works of Wilson [W1], [W3], [W4], [WF]. In retrospect one can say that the same *ideas* had been used several times in earlier works, for instance in (1), (2), and (5); furthermore, it would be very easy to illustrate the methods used in the analysis of the above problems without even mentioning the words *renormalization group*; this certainly happened in the early works on (1), (2) and (5).

where P is the Gaussian process with propagator

$$\overline{C}_{xy} = \frac{1}{(2\pi)^d} \int\limits_{-\infty}^{+\infty} \frac{\chi(p) e^{ip(x-y)}}{p^{2-\gamma}} d^d p \; , \qquad (2.13)$$

where $\tilde{J}(p)$ in (2.4) has been simplified into $p^{2-\gamma}$ and $\chi(p)$ is a cutoff function corresponding to the integral extremes $\pm\pi p_0$ in (2.8), and the integral has to be interpreted literally. For instance, a good choice — essentially as good as any other with the same features — is $\chi(p) = (p/p_0)^{2-\gamma} \int_1^\infty e^{-\alpha p^2/p_0^2} \alpha^{-\gamma/2} d\alpha$.

The fact that the volume is finite and that there are boundary conditions is important in some arguments: for example, when translation invariance is used or cancelations due to integrations by parts are exploited.

However, to avoid rather clumsy expressions, we shall use $\overline{C}$ in (2.11) or (2.13) as propagator rather than the periodized version C^1: in this way one breaks translation invariance, but we shall nevertheless perform the mentioned cancelations, where they are needed, without worrying about the inconsistency; it is possible to see that all the results reported in this review could be correctly reformulated in terms of the original model (2.7), (2.8). The reason is simply that all the rigorous results depend on bounds on the propagator (or better its multiscale decomposition, see chapter 5), which are uniform in the volume. The formal results are usually obtained in the literature, on the other hand, without worrying about "boundary corrections" and are very often given "in the limit $\Lambda \to \infty$."

The model (2.12), (2.13) is manifestly close to the problem of euclidean quantum field theory. The latter problem is, in fact, related to the analysis of the same integral with $\chi(p)$ replaced by $(1 - \chi(p))$ and $\gamma = 0$; more generally it is formulated as the problem of studying

$$\int P(d\psi) \exp \int_\Lambda \left(-\lambda \psi_x^4 - \mu \psi_x^2 \right) dx \; , \qquad (2.14)$$

with $P(d\psi)$ being a Gaussian measure with propagator given by (2.13) with $\overline{C}$ replaced by

$$\overline{C}_{xy} = \frac{1}{(2\pi)^d} \int d^d p \, \frac{1 - e^{-p^2/p_0^2}}{p^2} e^{ip(x-y)} \; . \qquad (2.15)$$

Here p_0 is some inverse length scale fixed in advance (*bare mass* of the theory).

The above model is called the φ^4-*model* and it can be naturally generalized to the $V(\varphi)$ *model,* where V is an arbitrary function, by replacing the integral over Λ in (2.14) by

$$- \int_\Lambda V(\psi_x)dx \ . \tag{2.16}$$

We have to make more precise the meaning of *the problem of studying the two integrals* (2.12) or (2.14).

The first is really a *problem* only in the limit $\Lambda \to \infty$, i.e., it is an *infrared* problem because its difficulty is due to the singularity at $p = 0$ of the integral in (2.13). It is usually formulated in terms of the singularities in the β-dependence of the *free energy*

$$F(\beta) = - \lim_{\Lambda \to \infty} \frac{1}{\beta |\Lambda|} \log Z(\beta) \ , \tag{2.17}$$

or in terms of the *generating functional* $S_\Lambda(f)$,:

$$\exp S_\Lambda^T(f) =$$
$$= \frac{1}{Z(\beta)} \int e^{\left[\psi(f) + \int_\Lambda \left(-L_0 \beta^{-2} \psi_x^4 - M_0(\beta)\beta^{-1}\psi_x^2\right)dx\right]} P(d\psi) \ , \tag{2.18}$$
$$\psi(f) \equiv \int \psi_x f(x)dx \ .$$

The equation (2.18) is usually written, with obvious symbols

$$S_\Lambda^T(f) = \log \langle e^{\psi(f)} \rangle \ . \tag{2.19}$$

The generating functional $S_\Lambda^T(f)$ is analytic in f and its Taylor coefficients

$$S_\Lambda^T(x_1 \cdots x_n) = \left. \frac{\delta^n}{\delta f(x_1) \cdots \delta f(x_n)} \log \langle e^{\psi(f)} \rangle \right|_{f=0} \tag{2.20}$$

are called the *truncated* Schwinger functions.

One is interested in $S^T(x_1, \cdots) = \lim_{\Lambda \to \infty} S_\Lambda^T(x_1, \cdots)$ and more precisely in their behavior when their arguments are sent far apart. For β small, $\beta < \beta_c$, the behavior of $S^T(x_1, x_2)$, for instance, is well known to be typically "the same as that" of[1] $J(x_1 - x_2)$, but for $\beta = \beta_c$ it may

---

[1] At high temperature and low density this is even a theorem in many cases; see, for example, [Gr], [BGr], and references therein.

change, signaling the (normal) *critical point*, to that of $C(x_1 - x_2)$ (i.e., from $1/|x_1 - x_2|^{d+2-\gamma}$, see (2.6), to $1/|x_1 - x_2|^{d-2+\gamma}$ which is slower because $\gamma < 2$, see (2.5)).

The statement "the same as that" of J is valid in the cases in which J is eventually monotonic. In the case that J has a decay faster than exponential, it means that the correlations decay exponentially. For J decaying slower than exponential, but faster than $|x|^{-d}$, it literally means asymptotic to J.

In contrast to the infrared problem (2.12), (2.13), the field theory problem (2.14), (2.15) (or (2.16)) is nontrivial already for Λ finite. In fact it is easy to see that (2.14) is not properly defined, if $d > 1$. The reason is simply that ψ_x^4 is not well defined, hence (2.14) itself is not meaningful: because if one chooses a field ψ randomly with probability distribution $P(d\psi)$ given by (2.15), then with probability 1 it is $\psi_x^2 = \psi_x^4 = \infty$. In other words, ψ_x^4 as well as $\int_\Lambda \psi_x^4$ is not a P-measurable function, hence it cannot be integrated by P (and its exponential either cannot be integrated or has a zero integral depending on whether $\int_\Lambda \left(-\lambda \psi_x^4 - \mu \psi_x^2 \right) dx$ is plus or minus infinity).

The proper definition of the (finite Λ) ultraviolet problem is therefore the following. Let N be a *cutoff* parameter and replace (2.15) by

$$\overline{C}_{xy}^{(\leq N)} = \frac{1}{(2\pi)^d} \int d^d p \, \frac{e^{-2^{-2N} p^2/p_0^2} - e^{-p^2/p_0^2}}{p^2} e^{ip(x-y)} \,, \qquad (2.21)$$

and let P_N be the Gaussian measure with propagator (2.21). Let $\lambda_N > 0$, μ_N real and define

$$S_N^T(f) = \log \frac{\int e^{\psi(f)} e^{-\int_\Lambda (\lambda_N \psi_x^4 + \mu_N \psi_x^2) dx} P_N(d\psi)}{\int e^{-\int_\Lambda (\lambda_N \psi_x^4 + \mu_N \psi_x^2) dx} P_N(d\psi)} \,, \qquad (2.22)$$

which is now a well-defined integral (in fact, the regularity of $C_{xy}^{(\leq N)}$ implies that ψ_x is a C^∞ function of x with probability 1 with respect to P_N).

We keep Λ finite and try to see whether one can find λ_N, μ_N so that the limit S^T, as $N \to \infty$, of $S_N^T(f)$ exists. To avoid triviality one has to require also that $S^T(f)$ shall not be a quadratic form in f. If $\lambda_N \to 0$ fast enough and $\mu_N \to 0$ fast enough then it is easy to prove that, in fact,

$$\lim_{N \to \infty} S_N^T(f) = \frac{1}{2} \int f(x) C_{xy} f(y) dx dy \equiv \frac{1}{2} (f, Cf) \,, \qquad (2.23)$$

obviously uninteresting and trivial, since the r.h.s. is the value of (2.22) when $\lambda_N = \mu_N = 0$.

More generally, field theory with an arbitrary V, see (2.16), can be formulated by replacing V by V_N, where V_N is allowed to vary with N in some space of functions (e.g., the sixth-degree polynomials) of ψ_x or, even more generally, in some space of function of ψ_x and $\partial\psi_x$. Then one studies the functions $S_N^T(f)$ defined by (2.23) with the integral in the exponent replaced by

$$\int_\Lambda V_N(\psi_x)dx \, , \qquad (2.24)$$

and one asks similar questions.

The latter problems really are the problem of trying to find a probability measure on the space of the fields ψ which is not Gaussian yet is *local*, i.e., obtained by modifying the Gaussian measure $P(d\psi)$ with propagator (2.15), by multiplying it by a function of the form $\exp - \int V(\psi_x)dx$. As such, the problem can be further generalized, but we shall not dwell on this (see [GR]). We shall discuss, in what follows, only the ultraviolet problem with V_N restricted to being a fourth-order polynomial in ψ_x and second-order in $\partial\psi_x$. This is what is called traditionally the φ^4-*problem* and it will be considered for $1 < d \leq 4$.

Chapter 3
Other Functional Integrals: Fermi Sphere and Bose Condensation

In the previous section we have considered two main examples of problems associated with functional integrals. There are many others. Here we mention two more problems: the theory of the ground state of a Fermi liquid, and that of a Bose gas. For lack of space we shall study only the very simplest cases: spinless fermions in one dimension and spinless bosons in three dimensions.

3.1 *The $d = 1$ Fermi Liquid*

This case is surprisingly rich in structure in spite of its absolute simplicity if compared with the more interesting case of one-dimensional spinning fermions or with the far more interesting $d > 1$ cases (see [BG2]).

The problem is the following. Let

$$H = \sum_{s=1}^{N} \left(-\frac{\Delta_{\vec{x}_i}}{2m} - \mu \right) + 2\lambda \sum_{i<j} v(\vec{x}_i - \vec{x}_j) \tag{3.1}$$

be the Hamiltonian describing a system of N fermions in $\mathbf{R}^d$, enclosed, as in chapter 2, in a periodic box Λ, interacting with a pair potential v that is supposed to be C^∞, and with short range.

Let $\psi_{\vec{x}}^{\pm}$ be the creation and the annihilation operators for the fermions and define, for $\beta \geq t_i \geq 0$, $i = 1, \ldots, n$, with $t_i \neq t_j$ for $i \neq j$:

$$S_{\sigma_1 \cdots \sigma_n}(\vec{x}_1 t_1, \ldots, \vec{x}_n t_n) = (-1)^\pi \lim_{\beta \to \infty} \lim_{\Lambda \to \infty} \cdot \tag{3.2}$$

$$\cdot \frac{\mathrm{Tr}\left(e^{-(\beta - t_{\pi(1)})H} \psi_{\vec{x}_{\pi(1)}}^{\sigma_{\pi(1)}} e^{-(t_{\pi(1)} - t_{\pi(2)})H} \psi_{\vec{x}_{\pi(2)}}^{\sigma_{\pi(2)}} \cdots \psi_{\vec{x}_{\pi(n)}}^{\sigma_{\pi(n)}} e^{-t_{\pi(n)}H} \right)}{\mathrm{Tr}\, e^{-\beta H}},$$

where $\sigma_i = \pm 1$, π is the permutation of $(1, \ldots, n)$, such that $t_{\pi(1)} > t_{\pi(2)} > \cdots > t_{\pi(n)}$ and $(-1)^\pi$ is the permutation parity. In particular, consider

$$S(x) = S_{-+}((\vec{x}, t), (\vec{0}, 0)) \qquad \text{if } x = (\vec{x}, t) \in \mathbf{R}^{d+1} . \tag{3.3}$$

The functions (3.2) and (3.3) describe the properties of the *ground state* of the above Fermi system (essentially *by definition*) in the grand canonical ensemble with chemical potential μ and when the mass of the particles is m. They are called the *Schwinger functions* of the ground state of the system (3.1).

The case $\lambda = 0$ is trivial and one finds that, if $\mu > 0$, $S(x) = \lim_{\beta \to \infty} \lim_{\Lambda \to \infty} g(x)$, with

$$g(x) = \sum_{\vec{n} \in \mathbf{Z}^d, n_0 \in \mathbf{Z}} (-1)^{n_0} \bar{g}(\vec{x} + \vec{n} l_\Lambda, t + n_0 \beta) \; ,$$

$$\bar{g}(x) = \frac{1}{(2\pi)^{d+1}} \int \frac{e^{-ikx}}{-ik_0 + (\vec{k}^2 - p_F^2)/2m} dk_0 d^d\vec{k}$$

$(k = (\vec{k}, k_0)$, $x = (\vec{x}, x_0)$ and $p_F = (2m\mu)^{1/2})$, which shows that the free-system ground state has a *one-particle* distribution function $S(x)$ singular on the "Fermi sphere" $k_0 = 0$, $|\vec{k}| = p_F$ [1]. In fact, the occupation number of the momentum $\vec{k}$ is related to the Fourier transform of $-g(\vec{x}, 0^-)$ which, in the limit $\beta \to \infty$, is

$$n_{\vec{k}} = \chi(|\vec{k}| < p_F) = \langle a_{\vec{k}}^+ a_{\vec{k}}^- \rangle \; ,$$

the familiar result (*Fermi distribution*).

The importance of the function (3.4), the *free propagator*, was discovered by Bloch and De Dominicis, who started a series of papers culminating in a rather final perturbation theory of the Fermi surface by Luttinger and Ward [BD], [LW].

The conclusion of the theory is that it is possible to express the interacting S-functions in (3.2) as a functional integral with propagator $g(x)$ described by (3.4). In fact, one finds, if $\Lambda = [-\frac{1}{2}\beta, \frac{1}{2}\beta] \times [-\frac{1}{2}L, \frac{1}{2}L]$ is considered with periodic boundary conditions,

$$S_{\sigma_1 \cdots \sigma_n}(x_1, \ldots, x_n) = \lim_{\beta \to \infty} \lim_{\Lambda \to \infty} \cdot$$

$$\frac{\int P(d\psi) \, e^{-\lambda \int_\Lambda \psi_{\bar{x}}^+ \psi_{\bar{x}}^- \psi_{\bar{y}}^+ \psi_{\bar{y}}^- \delta(x_0 - y_0) v(\vec{x} - \vec{y}) dx dy} \, \psi_{x_1}^{\sigma_1} \cdots \psi_{x_n}^{\sigma_n}}{\int P(d\psi) \exp \int_\Lambda \cdots \cdots} \; .$$

The main feature[2] of (3.6) is that the $\psi_x^\pm$ are not elements of the usual algebra of the real numbers, but they form, as x varies in $\mathbf{R}^{d+1}$, a basis for

---

[1] For a derivation of (3.4), see appendix 1.

[2] For a discussion of the notion of Grassmann algebra and of Grassmannian integration, see appendix 2.

a Grassmann algebra $\mathcal{G}$. The latter consists in the linear combination of monomials of the form $\psi_{x_1}^+ \cdots \psi_{x_n}^+ \psi_{y_1}^- \cdots \psi_{y_n}^-$; two such combinations are identified if one can be reduced to the other by assuming the identities

$$\psi_x^\sigma \psi_y^{\sigma'} + \psi_y^{\sigma'} \psi_x^\sigma \equiv 0, \qquad \forall x, y, \sigma, \sigma' . \qquad (3.7)$$

The $\int P(d\psi) \cdot$ is a linear functional over $\mathcal{G}$; thus it suffices to define the P-integral of the most general monomial. The latter is defined by the formula (*Wick's rule*)

$$\int P(d\psi) \, \psi_{x_1}^- \cdots \psi_{x_n}^- \psi_{y_1}^+ \cdots \psi_{y_n}^+ = \sum_\pi (-1)^\pi \prod_{i=1}^n g(x_i - y_{\pi(i)}) , \qquad (3.8)$$

the sum being over the $n!$ permutations π of $(1, \ldots, n)$ and $(-1)^\pi$ is the permutation parity. The propagator g is as described in (3.4).

The derivation of (3.6) from (3.2), (3.8) is a matter of simple algebra based on Trotter's product formula (see [LW], [BG1]); it is checked order by order of the λ expansion of both sides of (3.6). If the various contributions of the same order in λ are collected together and summed, the series resulting from this procedure are absolutely convergent (see [GK1], [BGPS]), and one says that the integral in (3.6) is well defined and given by the appropriate sum (the difficult problem is of course to prove bounds uniform in Λ).

3.2 *The $d = 3$ Bose Gas.*

The Hamiltonian is the same as (3.1), acting, however, on symmetric wave functions.

Let $\varphi_x^\pm$ be the creation and annihilation operators for bosons, and define the finite L positive β, Schwinger functions $S_{\sigma_1,\ldots,\sigma_n}^{\beta,L}(x_1,\ldots,x_n)$ by the ratio in the r.h.s. of (3.2), with $\psi^\pm \to \varphi^\pm$.

The $\lambda = 0$ case is, of course, trivial, and the functions $S^{\beta,L}$ are given by the Wick rule (3.8) with $(-1)^\pi \to 1, \psi^\pm \to \varphi^\pm$ and $g(x)$ replaced by

$$S^{\beta,L}(x) = S_{-+}^{\beta,L}(x,0) =$$

$$L^{-d} \sum_{\vec{k}} e^{-i\vec{k}\cdot\vec{x}} e^{-\varepsilon(\vec{k})t} \left(\frac{\vartheta(t>0)}{1 - e^{-\beta\varepsilon(\vec{k})}} + \frac{e^{-\beta\varepsilon(\vec{k})}\vartheta(t\leq 0)}{1 - e^{-\beta\varepsilon(\vec{k})}} \right) = \qquad (3.9)$$

$$= \rho e^{\mu t} \left(e^{-\beta\mu}\vartheta(t>0) + \vartheta(t\leq 0) \right) + L^{-d} \sum_{\vec{k}\neq\vec{0}} (\text{same as above}) ,$$

where $\varepsilon(\vec{k}) = \frac{\vec{k}^2}{2m} - \mu$, $\vartheta(x > 0)$ means 1 if $x > 0$ and 0 otherwise, and ρ is defined by

$$\rho = L^{-d} e^{\beta\mu} (1 - e^{\beta\mu})^{-1}, \qquad \beta\mu = \log \frac{\rho L^d}{1 + \rho L^d}, \qquad (3.10)$$

and it will be called the *condensate density* (equal to the actual total density $S^{\beta,L}(\vec{0}, 0^-)$ in the free case).

One finds, if $\rho > 0$ is kept fixed,

$$S^{\beta,L}(x) \xrightarrow[\beta,L\to\infty]{} S(x) = S_\rho(x) \equiv \rho + \frac{1}{(2\pi)^4} \int dk \, \frac{e^{-ikx}}{-ik_0 + \vec{k}^2/2m} \, . \qquad (3.11)$$

Note that $\mu \to 0^-$ as $\beta, L \to \infty$: in fact, one cannot fix $\mu = 0$ (because one would divide by 0 in (3.9)), nor $\mu < 0$ (because this would imply $S(\vec{x}, t) = 0$ for $t < 0$, i.e. zero density: *the vacuum*). The choice $\mu > 0$ is not allowed; in fact, even if one took L such that $\varepsilon(\vec{k}) \neq 0$ for all $\vec{k}$'s (thus avoiding a division by 0), one could not avoid generating an unphysical negative density.

Following the same procedure used to deduce (3.6) from (3.2), (3.8), it can be easily found that (3.2) can be written as

$$S_{\sigma_1\dots\sigma_n}(x_1\dots x_n) = \lim_{\Lambda\to\infty} \int \frac{\varphi_{x_1}^{\sigma_1}\dots\varphi_{x_n}^{\sigma_n} e^{-V(\varphi)} P(d\varphi)}{\int e^{-V(\varphi)} P(d\varphi)} \, , \qquad (3.12)$$

where $\Lambda = [-\frac{1}{2}\beta, \frac{1}{2}\beta] \times [-\frac{1}{2}L, \frac{1}{2}L]^3$, and $P(d\varphi)$ is a Gaussian integral, such that

$$\int \varphi_x^- \varphi_y^+ P(d\varphi) = S_{\beta,L}(x - y) \, , \qquad \int \varphi_x^\sigma \varphi_y^\sigma P(d\varphi) = 0 \, , \qquad (3.13)$$

and,[3] if $x = (x_0, \vec{x})$, $y = (y_0, \vec{y})$,

$$V(\varphi) = \lambda \int_\Lambda v(\vec{x} - \vec{y})\delta(x_0 - y_0)\varphi_x^+ \varphi_x^- \varphi_y^+ \varphi_y^- \, dx \, dy \, . \qquad (3.14)$$

---

[3] The term "Gaussian distribution" attributed to $P(d\varphi)$ is justified because $\varphi_x^\pm$ can be regarded as complex Gaussian random fields, defined in terms of two real, independent, Gaussian random fields Φ_x^1, Φ_x^2 and two real Gaussian, scalar variables ξ^1, ξ^2. One can set

$$\varphi_x^\pm = \xi^1 \pm i\xi^2 + O^\pm(\Phi^1 \pm i\Phi^2) \, ,$$

where $O^+ = 1$ and $O^- = -\partial_t - (2m)^{-1}\Delta$. Then, if the covariance of ξ^j is $\rho/2$ and the Fourier transform of the covariance of Φ^σ is $[k_0^2 + (\vec{k}^2/2m)^2]^{-1}$, the propagator for the fields $\varphi_x^\pm$ is (3.11). Note that, in this representation, the fields φ^+ and φ^- are not complex conjugate. The asymmetry of the above representation implies the existence

This time the integrals in (3.12) are true functional integrals, convergent if $\lambda v(\vec{x} - \vec{y}) > 0$ (so that $V_\Lambda(\varphi) \geq 0$, because φ_x^- and φ_x^+ are complex conjugate).

Remarks

1. Note that, in the case of the Fermi gas, formula (3.6) can be written in several ways; if, instead of setting $\mu = \frac{p_F^2}{2m}$, one sets $\mu = \frac{p_F^2}{2m} - \nu$, then the Schwinger functions are given by

$$S_{\sigma_1 \ldots \sigma_n}(x_1 \ldots x_n) = \lim_{\Lambda \to \infty} \int \psi_{x_1}^{\sigma_1} \ldots \psi_{x_n}^{\sigma_n} \frac{e^{-V(\psi)} P(d\psi)}{\int e^{-V(\psi)} P(d\psi)} \, ,$$

$$V(\psi) = \lambda \int v(\vec{x} - \vec{y}) \delta(x_0 - y_0) \psi_x^+ \psi_x^- \psi_y^+ \psi_y^- \, dx \, dy + \qquad (3.15)$$

$$+ \nu \int \psi_x^+ \psi_x^- \, dx \, ,$$

where P is the measure with propagator $\bar{g}(x)$, see (3.4).

In other words the chemical potential can be arbitrarily split into a part, $\frac{p_F^2}{2m}$, that is regarded as "part of the free system" and into a "part of the interaction." Hence p_F is not intrinsically defined by (3.15).

2. The same comment can be made about the Bose gas: the chemical potential can be written $\mu = \frac{1}{\beta} \log L^d \rho (1 + L^d \rho)^{-1} - \nu$, and (3.12) becomes (3.15) with $\psi \to \varphi$. This means that we can fix arbitrarily ρ provided we compensate it with a value ν such that μ stays the same. Hence ρ is not intrinsically defined by (3.15).

3. Nevertheless one can define uniquely the notions of Fermi momentum and Bose condensate density in terms of properties of the Schwinger functions. One notes that, in the Fermi gas with $\lambda = 0$ and given p_F, the pair Schwinger function $S(\vec{x}, 0^-)$ has asymptotic behavior that is $\propto \frac{\sin p_F |\vec{x}|}{|\vec{x}|}$. Therefore, one says that an interacting Fermi gas has Fermi momentum p_F if $S(\vec{x}, 0^-)$ oscillates as $|\vec{x}| \to \infty$ as $\sin p_F |\vec{x}|$. Of course

---

of another representation with the $O^\pm$ operators exchanged; but no representation is possible with φ^+ conjugate of φ^-. This makes the functional integral somewhat unusual and it is perhaps better to regard it as a formal rule for the calculation of the perturbation expansions; this causes no problems until one starts worrying about the remainder estimates. We are not aware of any attempt to treat such question with mathematical accuracy.

if one uses (3.15) with $\mu = \frac{p_F^2}{2m} + \nu$, the function $S(\vec{x}, 0^-)$ will neither oscillate with scale p_F nor with scale $\sqrt{2m\mu}$; but it will oscillate with some scale $p_F(\lambda, \mu)$ (assuming that things are not too different in the interacting and in the free cases, for small coupling at least).

This shows that a natural approach would be to fix a priori p_F as the scale of the oscillations of the interacting pair Schwinger function, and to see if there is a value of the chemical potential μ such that, writing $\mu = \frac{p_F^2}{2m} - \nu$, the (3.15) produces a $S(\vec{x}, 0^-)$ oscillating exactly on scale p_F.

This means that one regards p_F as a "physical constant" and μ as a "bare constant" to be fixed to generate a model whose physical Fermi momentum is the prescribed one. One expects that ν will turn out to be $\nu = \nu(p_F, \lambda) = O(\lambda)$, i.e. $\mu = \frac{p_F^2}{2m} - \nu(p_F, \lambda)$.[4]

One expects that the perturbation theory would be simpler at fixed p_F rather than at fixed μ, because in this way the Schwinger functions have the same singularities as the free ones, or at least some common singularities, so that they can be more naturally regarded as perturbations of each other.

4. Likewise in the Bose gas one sees that the asymptotic behavior of the free $S(x)$ as $x \to \infty$ is simply $S(x) \to \rho$ (see (3.11)). But again, if one uses (3.15) with $\psi \to \varphi$ and with P having a propagator (3.11) and ν fixed, we can expect, at best, that $S(x) \xrightarrow[x \to \infty]{} \rho + r(\nu, \lambda)$. Therefore, we expect that a less singular perturbation analysis will be necessary, if ρ is fixed a priori (and called the *condensate density*), and $\nu = \nu(\rho, \lambda)$ in (3.15) is determined so that $S(x) \xrightarrow[x \to \infty]{} \rho$ exactly. We see that the Bose and Fermi gases are completely analogous in this respect.

5. Once the physical parameters p_F or ρ are fixed in (3.15) by suitably determining ν, one can inquire whether the subleading corrections to the asymptotic behavior are also in agreement with the (respective) free-gas cases.

The subleading behaviors are

$$S_{fermi,\,d=1}(x) \sim -\frac{1}{\pi} \frac{\sin(p_F x - \arctan v_F t/x)}{(x^2 + (v_F t)^2)^{1/2}} \,, \qquad v_F = \frac{p_F}{m} \,,$$

$$S_{bose,\,d=3} - \rho \sim \frac{e^{-m\vec{x}^2/t}}{(\pi t m^{-1})^{3/2}} \,, \tag{3.16}$$

---

[4] If one really insists in fixing μ rather than p_F, then the last relation should be regarded as an equation for p_F, given μ, λ, and thus it would determine p_F.

so that, if one could prove that the above are the correct asymptotic behaviors also in the interacting cases, for suitably chosen $m = m(\lambda, p_F)$ or $m = m(\lambda, \rho)$, it would be natural to use (3.16) as definitions of the *physical mass* of the particles. Or, as in the cases of p_F and ρ, one could fix the physical mass m and use (3.15) with a free field with propagator containing the physical value of the mass parameter m and add to the V_Λ in (3.15) a term

$$\alpha \int_\Lambda \frac{-\Delta - p_F^2}{2m} \psi_x^+ \psi_x^- \, dx \ , \tag{3.17}$$

with $\alpha(m, p_F, \lambda)$ such that the interacting Schwinger function has the behavior (3.16). The Bose case is essentially identical, with $\psi \to \varphi$.

6. *However* it may happen that the subleading behaviors *do not* behave as (3.16). For instance, in the Fermi gas case it turns out that the asymptotic behavior is

$$\frac{S^0(x)}{(x^2 + (v_F t)^2)^{\eta(\lambda)/2}} \tag{3.18}$$

if $S^0(x)$ denotes the free Schwinger function, and if η is a suitably defined function $\eta(\lambda) = O(\lambda^2)$. This still defines the parameter v_F, hence the physical mass p_F/v_F. We discuss this point in chapter 11.

In the Bose case also one does not expect (3.16) to hold. We shall discuss this point in chapters 13 and 14. The result will be essentially the same: the physical mass can be fixed by adding to V_Λ in (3.15) a term like (3.17) with $\psi \to \varphi$.

The existence of substantial corrections to the (3.16) (such as (3.18)) is usually called an *anomaly* to stress that the asymptotic behavior of the Schwinger functions is *not* the free one.

Chapter 4
Effective Potentials and Schwinger Functions

The problems of Chapters 2 and 3 are studied by the *renormalization group approach* via the theory of the *effective potentials*.

There are various objects that bear the name *effective potential*. Our definition is one variant of the concept introduced in the basic paper of Jona-Lasinio[J] (see also [CW], [Po1]).

Consider the functional integrals (2.12), (2.14), (3.6) and define

$$e^{-V_{\text{eff}}(\varphi)} \equiv \frac{1}{\mathcal{N}} \int P(d\psi) e^{-V(\psi+\varphi)} , \qquad (4.1)$$

where $\mathcal{N}$ is a normalization constant, chosen so that $V_{\text{eff}}(0) = 0$, and, respectively,

$$
\begin{aligned}
V(\psi) &= \int_{\Lambda} \left(L_0 \beta^{-2} \psi_x^4 + M_0(\beta)\beta^{-1}\psi_x^2 \right) dx , \\
V(\psi) &= \int_{\Lambda} \left(\lambda_N \psi_x^4 + \mu_N \psi_x^2 \right) dx , \qquad (4.2) \\
V(\psi) &= \lambda \int_{\Lambda} v(\vec{x} - \vec{y})\delta(x^0 - y^0)\psi_x^+ \psi_x^- \psi_y^+ \psi_y^- \, dx dy + \nu \int_{\Lambda} \psi_x^+ \psi_x^- dx ,
\end{aligned}
$$

and $P(d\psi)$ denotes the Gaussian measure with propagator (2.13), (2.21) or the Grassmannian Gaussian defined by (3.8), (3.4). The field φ is a test function (i.e., a smooth, rapidly decreasing function on $\mathbf{R}^d$) in the first two cases. In the third case φ is, instead, an element of a Grassmann algebra obtained by enlarging the basic one, generated by the $\psi_x^{\pm}$ field, by adding to it new basic elements $\varphi_x^{\pm}$ (which, therefore, anticommute with each other as well as with the ψ's).

The case of the Bose condensation problem is very analogous to the first two (see chapter 13).

The effective potential (4.1) is the object of main interest. It might look somewhat unnatural; we have, in fact (see chapters 2, 3), stated that what is really interesting is the family of functions S^T (see (2.20)). It is, however, easy to see that there is a very simple connection between the effective potential and the truncated Schwinger functions S^T. In the *scalar cases* one has

$$S^T(f) = \frac{1}{2}(Cf, f) - V_{\text{eff}}(Cf) \ . \tag{4.3}$$

This means that the convolution of the effective potential with the free propagator is the correction that has to be added to the free Schwinger functions (given, trivially, by $S_0^T(f) \equiv \frac{1}{2}(Cf, f)$; see (2.23)) to obtain the ones in presence of interaction.

More explicitly (4.3) says, for $n > 2$,

$$S^T(x_1, \ldots, x_n) = -\int V_{\text{eff}}(y_1 \cdots y_n) C_{x_1 y_1} \cdots C_{x_n y_n} dy_1 \cdots dy_n \ , \tag{4.4}$$

if $S^T(x_1 \cdots)$, and $V_{\text{eff}}(y_1 \cdots)$ are the Taylor coefficients of $S^T(f)$, $V_{\text{eff}}(f)$.

For $n = 2$ and if $\hat{S}^T(k)$ denotes the Fourier transform of $S^T(x_1, x_2)$ in $x_2 - x_1$, and $\hat{V}_{\text{eff}}(k)$ denotes the analogous transform of $V_{\text{eff}}(x_1, x_2)$ (assuming $\Lambda \to \infty$),

$$\hat{S}^T(k) = \hat{C}(k) - \hat{C}(k)^2 \hat{V}_{\text{eff}}(k) \ . \tag{4.5}$$

(4.5) does not look like a double convolution, like (4.4), because we can exploit the translation invariance. The case of the Bose condensation problem is the same.

In the case of Fermi liquids there is a similar formula. On can define

$$S^T(\varphi) = \log \int P(d\psi) \ e^{-V(\psi) + \int (\psi_x^+ \varphi_x^- + \varphi_x^+ \psi_x^-) dx} \tag{4.6}$$

and the connection between $S^T(x_1 \cdots x_n)$ (where n is even and the first $n/2$ x-variables are associated with ψ^- fields) and $V_{\text{eff}}(x_1 \cdots x_n)$ is still (4.4), (4.5).

We just give a formal proof of (4.3), which could be made rigorous (as long as Λ is kept finite). In fact, the basic reason for the validity of (4.4),(4.5), i.e., of (4.3), is the important formal representation of $P(d\psi)$ as

$$P(d\psi) \ \text{``}=\text{''} \ \text{const} \ e^{-\frac{1}{2}(C^{-1}\psi, \psi)} \prod_{x \in \mathbf{R}^d} d\psi_x \ . \tag{4.7}$$

Starting from (4.7), one sees that (redefining ψ as $\psi' - \varphi$)

$$e^{-V_{\text{eff}}(\varphi)} = \text{const} \int e^{-\frac{1}{2}(C^{-1}\psi, \psi) - V(\varphi + \psi)} \prod d\psi_x =$$

$$= \text{const} \int e^{-\frac{1}{2}(C^{-1}\psi', \psi') - \frac{1}{2}(C^{-1}\varphi, \varphi) + (C^{-1}\varphi, \psi')} e^{-V(\psi')} \prod d\psi_x' = \tag{4.8}$$

$$= e^{-\frac{1}{2}(C^{-1}\varphi, \varphi) + S^T(C^{-1}\varphi)} \ ,$$

yielding (4.3).

The case of fermion liquids is treated similarly by using the analogue of (4.7),

$$P(d\psi) \text{ "="} \text{const } e^{-(g^{-1}\psi,\psi)} \prod_{x \in \mathbf{R}^{d+1}} d\psi_x , \tag{4.9}$$

where $\prod_x d\psi_x$ is formally defined as a Grassmannian integration rule in which the $g(x-y)$, in (3.8), is replaced by $\delta(x-y)$. We use the notation

$$(g^{-1}\psi, \psi) = \int g^{-1}(x-y)\psi_y^+ \psi_x^- d^{d+1}x d^{d+1}y . \tag{4.10}$$

The representations (4.7), (4.9) are extremely useful for heuristic purposes. To turn the heuristic arguments into rigorous ones it is necessary, for instance, to imagine (4.7), (4.9) as the expressions obtained by replacing $\mathbf{R}^{d+1}$ by a lattice with small but positive lattice spacing and finite spatial extension, and then by considering the limit in which the lattice becomes $\mathbf{R}^{d+1}$. Or one can use the ideas introduced in appendix 2.

Chapter 5
Multiscale Decomposition of Propagators and Fields: Running Effective Potentials

All the model problems that we are considering have a common feature. When formulated as functional integrals, they give rise to propagators with some kind of singularity. The singularity is at $p = 0$ in the critical point theory of chapter 2, at $p = \infty$ in the ultraviolet problems of field theory considered in chapter 2, and both at $k_0 = 0$, $|\vec{k}| = p_F$ ($\vec{k} = 0$) *and* at $k = \infty$ in the Fermi liquid (Bose gas) case.

The Bose gas problem has some peculiarity that we shall discuss in chapter 13; here we shall consider only the other three problems.

The first two problems are simpler because the singularity occurs *just in one point*. The third problem has a singularity in one point (∞) plus a new type of singularity at the Fermi surface, which has codimension 2 (whatever d is). To be more precise, the ultraviolet problem in its regularized version involves a *non-singular* propagator (2.21), which, however, becomes singular at $p = \infty$ when one considers, as one must, the limit $N \to \infty$. The Fermi liquid problem has two *problems* built in: an ultraviolet problem and an infrared one (and the singularity at the Fermi surface manifests itself in a slowly oscillating decay, at large distance, of the propagator).

Since the novelty of the third case rests on the singularity on the Fermi surface we shall, from now on, simplify also this problem by replacing $\bar{g}$ in (3.4) by:

$$\bar{g}(x) = \frac{1}{(2\pi)^{d+1}} \int \frac{e^{-(k_0^2 + \varepsilon(\vec{k})^2)/p_0^2} e^{-ikx}}{-ik_0 + \varepsilon(\vec{k})} d^{d+1}k \,, \tag{5.1}$$

where $\varepsilon(\vec{k}) = (\vec{k}^2 - p_F^2)/2m$, thus eliminating the singularity at ∞, by cutting off the k-values large compared to some prefixed scale p_0, which can be conveniently taken to be the inverse of the range of the interaction potential.

Moreover, we shall neglect the boundary problems related to the finite volume also in the Fermi case (see discussion in chapter 2); hence, we shall take (5.1) as the full covariance.

The singularities present in the propagators (2.13) (infrared singular at $p = 0$), (2.22) (ultraviolet singular at $p = \infty$), and (5.1) (singular on a surface of codimension 2 away from the origin) can be "resolved" into scales by using a procedure well known in harmonic analysis (in the theory of convergence of Fourier series (see [C], [F]), and, more generally, of expansions in orthogonal functions, like the Walsh series (see [Go])).

One simply writes the identities for, respectively, (2.13), (2.21), and (5.1):

$$\frac{\chi(p)}{p^{2-\gamma}} = \frac{1}{p_0^{2-\gamma}} \sum_{h=-\infty}^{0} \Delta_{2-\gamma}\left(2^{-h}p/p_0\right) 2^{-(2-\gamma)h} \ .$$

$$\frac{e^{-2^{-2N}p^2/p_0^2} - e^{-p^2/p_0^2}}{p^2} = \frac{1}{p_0^2} \sum_{h=1}^{N} \Delta_2\left(2^{-h}p/p_0\right) 2^{-2h} \ .$$

$$\frac{e^{-(k_0^2+\varepsilon(\vec{k})^2)/p_0^2}}{-ik_0 + \varepsilon(\vec{k})} = \frac{1}{p_0^2} \sum_{h=-\infty}^{0} 2^{-2h}\left(ik_0 + \varepsilon(\vec{k})\right) \cdot$$
$$\cdot \int_1^4 e^{-\alpha\, 2^{-2h}(k_0^2+\varepsilon(\vec{k})^2)/p_0^2} d\alpha \ ,$$

(5.2)

where the $\Delta_{2-\gamma}$ are analytic functions of the square of their argument, with domain containing a strip along the real axis, and are rapidly decaying at ∞.[1]

The first two decompositions in (5.2) can be written as

$$\overline{C}_{xy} = \sum_{h=-\infty}^{0} \tilde{C}\left(2^h p_0(x-y)\right) 2^{(d-2+\gamma)h} \ ,$$

$$\overline{C}_{xy} = \sum_{h=1}^{N} \tilde{C}\left(2^h p_0(x-y)\right) 2^{(d-2)h} \ ,$$

(5.4)

where $\tilde{C}$ is a C^∞ function rapidly decreasing at ∞ and *independent* of h. In the infrared case, p_0^{-1} can be interpreted as a lattice spacing, while in the ultraviolet case p_0 can be thought of as a physical mass.

---

[1] *Exercise:* Show that

$$\Delta_{2-\gamma}(q) = \int_1^4 e^{-\alpha q^2} \frac{d\alpha}{\alpha^{\gamma/2}}$$

(5.3)

if $\chi(p)$ is chosen as after (2.13).

The last of (5.2) cannot be written simply as (5.4), i.e., as the sum of terms identical up to scaling. If $d = 1$, one can, however, see that ([BG1], appendix A), choosing units so that the Fermi velocity $p_F/m = 1$,

$$\bar{g}(x - y) = \sum_{h=-\infty}^{0} \sum_{\vec{\omega}=\pm 1} e^{-ip_F \vec{\omega}\cdot(\vec{x}-\vec{y})} 2^h \, \tilde{g}^{(h)} \left(2^h p_0(x - y), \vec{\omega}\right) , \tag{5.5}$$

$$\hat{\tilde{g}}^{(h)}(k, \vec{\omega}) = (ik_0 + \vec{\omega}\cdot\vec{k}) p_0 \Delta_2(k) \left(1 + o(2^h)\right) .$$

p_0^{-1} can be chosen as the range of the interaction to avoid the introduction of an extra-length scale.

As a final simplification of the model, we shall neglect the terms $o(2^h)$, so that $\tilde{g}^{(h)}$ becomes in fact h-independent and our fermion propagator takes the form

$$\bar{g}(x - y) = \sum_{\vec{\omega}=\pm 1} e^{-ip_F \vec{\omega}\cdot(\vec{x}-\vec{y})} \sum_{h=-\infty}^{0} 2^h \, \tilde{g} \left(2^h p_0(x - y), \vec{\omega}\right) , \tag{5.6}$$

$$\hat{\tilde{g}}(k, \vec{\omega}) = (ik_0 + \vec{\omega}\cdot\vec{k}) \tilde{C}(k) ,$$

and (5.4), (5.6) show manifest analogies. It is interesting to remark that (5.6), i.e., the decomposition of the leading part (as one approaches the Fermi surface or, in other terms, as $h \to -\infty$) of (3.4), corresponds to the approximation

$$\bar{g}(x - y) = \sum_{\vec{\omega}=\pm 1} \int \frac{e^{-ik(x-y)}}{-ik_0 + \vec{\omega}\cdot\vec{k}} e^{-ip_F \vec{\omega}\cdot(\vec{x}-\vec{y})} d^2k \tag{5.7}$$

if $d = 1$.

The above decompositions of the propagators can be used to represent the fields ψ_x as sums of other fields, with the integrations over the ψ's replaced by independent integrations over auxiliary fields. The representation will be called a *multiscale decomposition*.

In the scalar cases we write, respectively,

$$\psi_x = \sum_{h=-\infty}^{0} \psi_x^{(h)} \qquad \text{or} \qquad \psi_x = \sum_{h=1}^{\infty} \psi_x^{(h)} , \tag{5.8}$$

where $\psi_x^{(h)}$ are Gaussian fields with propagators

$$\delta_{hh'} \tilde{C} \left(2^h p_0(x - y)\right) 2^{(d-2+\gamma)h} . \tag{5.9}$$

So that the distribution of $\psi_x^{(h)}$ is the same as that of $\psi_x^{(0)}$ suitably rescaled, i.e.,

$$\psi_x^{(h)} \overset{\text{dist}}{=} 2^{(d-2+\gamma)h/2} \psi_{2^h x}^{(0)} \,. \tag{5.10}$$

And we can write the functional integral for the effective potential as

$$e^{-V_{\text{eff}}(\varphi)} = \frac{1}{N} \int \prod_h P(d\psi^{(h)}) \, e^{-V\left(\varphi + \sum_h \psi^{(h)}\right)} \,, \tag{5.11}$$

where h ranges between $-\infty$ and 0 in the infrared problems, and between 1 and N in the ultraviolet problems; V is given by the first two of (4.2), respectively.

Likewise, in the Fermi liquid case, equation (5.6) can be used, supposing for simplicity that $p_F L/\pi$ is integer, as follows. Let $\psi_{x\bar\omega}^{(0)}, \psi_{x\bar\omega}^{(-1)}$, ... be a basis for a Grassmann algebra and introduce the integration with the propagator

$$\delta_{\bar\omega\bar\omega'} \delta_{hh'} \tilde g \left(2^h p_0 (x-y), \bar\omega\right) 2^h \,. \tag{5.12}$$

Then, if we define

$$\psi_x^{\pm} = \sum_{\bar\omega = \pm 1} \sum_{h=-\infty}^{0} e^{\pm i p_F \bar\omega \cdot \bar x} \psi_{x\bar\omega}^{\pm(h)} \,, \tag{5.13}$$

we see that the effective potential can be written exactly as (5.11), with now $\varphi_{x\bar\omega}, \psi_{x\bar\omega}^{(h)}$ being interpreted as the basic fields of a Grassmann algebras and with V given by the last of (4.2). They will be called the *quasi particle* fields.

We can illustrate the interest of the multiscale decomposition by taking, for example, the ultraviolet case. The ψ_x's have p_0^{-1} as independence scale in the sense that $C_{xy} \sim 0$ if $|x-y|p_0 \gg 1$ so that we would like to regard our fields ψ_x's as constant on square boxes of size p_0^{-1}. This is, however, clearly impossible, because as x' tends to x, we have that $C_{xx'}$ diverges as $1/|x-x'|^{d-2}$.

But in the case of the $\psi_x^{(h)}$'s, the independence scale is $2^{-h}p_0^{-1}$ (i.e., $C_{xy}^{(h)} \sim 0$, if $2^h p_0^{-1}|x-y| \gg 1$), and it is now possible to regard the $\psi_x^{(h)}$'s as almost constant on boxes of size $2^{-h}p_0^{-1}$, because when x' tends to x it is $C_{xx'}^{(h)} \sim \tilde C(0) 2^{(d-2)h}$, i.e., $C_{xx'}^{(h)}$ becomes "essentially" constant when x, x' are in the same box of a lattice with lattice spacing $2^{-h}p_0^{-1}$.

Note that if one *defines* $C_{xx'}^{(h)} = 2^{(d-2)h}$, if x, x' are in the same box of a hierarchical sequence of lattices of size $2^{-h}p_0^{-1}$, and $C_{xy}^{(h)}$ is set equal to zero, if x, y belong to different boxes, one obtains what is called the *hierarchical model*, leading to the *Wilson recursion relation* (see [W3],[G2]).

The (5.11) leads naturally to the definition of *effective potential on scale h*,

$$e^{-V^{(h)}(\varphi)} = \int \prod_{h'>h} P(d\psi^{(h')}) e^{-V(\varphi + \sum_{h'>h} \psi^{(h')})} \,, \qquad (5.14)$$

so that $V_{\text{eff}}(\varphi) = \lim_{h \to -\infty} V^{(h)}(\varphi)$ in the infrared cases and $V_{\text{eff}}(\varphi) = \lim_{N \to \infty} V^{(0)}(\varphi)$ in the ultraviolet cases.

Hence the following recursion relation between the $V^{(h)}$ holds, by (5.10), (5.13), (5.14),

$$e^{-V^{(h-1)}(\varphi)} = \int P(d\psi^{(0)}) \, e^{-V^{(h)}(\varphi + 2^{\delta h} \psi^{(0)}_{2h.})} \,, \qquad (5.15)$$

where $\delta = (d - 2 + \gamma)/2$ in the scalar cases and $\delta = 1/2$ in the fermion case. The quantity δ is called the *scaling dimension* of the field.

The sequence $V^{(h)}$ as h varies in its range (i.e. $h = 0, -1, -2, \ldots$ in the infrared cases and $h = 1, 2, 3, \ldots, N$ in the ultraviolet cases) is called the *renormalization group flow*. Its analysis becomes the primary subject of investigation.

Sometimes the sequence $V^{(h)}$ is called the *running effective potentials* sequence (see [G2], [BG1]).

Chapter 6
Renormalization Group:
Relevant and Irrelevant Components
of the Effective Potentials

If we introduce the *dimensionless* effective potential (see (5.15))

$$\overline{V}^{(h)}(\varphi) \equiv V^{(h)}(2^{\delta h}\varphi_{2^h}.) \, ,\tag{6.1}$$

the recursion relation (5.15) becomes

$$e^{-\overline{V}^{(h-1)}(\varphi)} = \int P(d\overline{\psi})e^{-\overline{V}^{(h)}(2^{-\delta}\varphi_{2^{-1}.}+\overline{\psi})} \, ,\tag{6.2}$$

where P is the Gaussian measure describing the same distribution as that of the field $\overline{\psi}\equiv\psi^{(0)}$ (or in the fermionic case $\overline{\psi}\equiv\psi^{(0)}$).

In (6.2) we can try to forget the scale index h and the fact that the initial choice of $\overline{V}$ has a special form, and regard it as a map $\mathcal{R}$:

$$V'(\varphi) = -\log \int P(d\overline{\psi})e^{-V(2^{-\delta}\varphi_{2^{-1}.}+\overline{\psi})} \equiv \mathcal{R}V(\varphi) \, .\tag{6.3}$$

One should keep in mind that $\mathcal{R}$ is a noninvertible map, so at each application of $\mathcal{R}$ we are losing information, but we expect that what is kept is the essence of the problem, as will appear more clearly at the end (see also the Introduction).

Note that, had we taken correctly into account the volume dependence of the fields (see discussion in chapter 2), the measure $P(d\overline{\psi})$ and the map $\mathcal{R}$ would have a slight dependence on h, but the following analysis would not change in an essential way.

In trying to assign to $\mathcal{R}$ a domain of definition, one supposes that $V(\varphi)$ has the form, in the scalar cases,

$$V(\varphi) = \sum_{n,p_1...p_n} \int_{\Lambda} V_{n,p_1...p_n}(x_1 \ldots x_n)\Phi_{x_1}^{p_1} \ldots \Phi_{x_n}^{p_n} dx_1 \ldots dx_n \, .\tag{6.4}$$

The $V_{n,p_1...p_n}$ are suitable kernels (smooth up to some delta function factor) and Φ_x are fields ψ_x or $\partial_i\psi_x$ (in the latter case V_n should also contain the indices necessary to contract the gradient indices, omitted in (6.4) for simplicity of notation).

In the Fermi liquid problem, the (6.4) takes the form

$$V(\varphi) = \sum_\alpha \int_\Lambda V_\alpha(x_1 \ldots x_n) \Phi_{x_1} \cdots \Phi_{x_n} e^{i p_F \sum_i \sigma_i \bar{\omega}_i \bar{x}_i} dx_1 \ldots dx_n \ , \quad (6.5)$$

where $\Phi_{x_i} = \varphi^{\sigma_i}_{x_i \bar{\omega}_i}$ or $D \varphi^{\sigma_i}_{x_i \bar{\omega}_i}$, D being a convenient differential operator (which could be the gradient, but which will turn out, below, to be more conveniently chosen otherwise), the summation is over the labels $\alpha = (n, \omega_1, \ldots, \omega_n, \sigma_1, \ldots, \sigma_n)$, and the monomial in the Φ contains as many φ^+ as φ^- quasi-particle variables. The sum over the $\bar{\omega}$ indices should be interpreted as an integral, if $d > 1$, but we shall consider in the following only the case $d = 1$.

The integral in (6.4) has to be understood in the sense in which one usually understands the Hamiltonians of infinite systems. In principle, the integral in (6.4) has to be restricted to a finite region Λ, then one evaluates (6.3) with such a modified V. One gets a V'_Λ, which will be written, if possible, as (6.4) with kernels $V'^{(\Lambda)}_{n, p_1 \ldots p_n}$. Finally, the kernels $V'_{n, p_1 \ldots p_n}$ of V' will be the limits of the corresponding ones in $V'^{(\Lambda)}$.

If one performs this step (i.e., forgetting the finiteness of Λ and β and the related boundary conditions problems; see chapter 2), essentially amounting at a not yet justified interchange of limits, it is then not difficult to show that the above procedure leads to formal expressions for the V'_n in power series of the V_n whenever the V_n verify certain properties.

Such properties can be checked to hold for all the V's that are obtained by successive applications of $\mathcal{R}$ to the V's that we have considered in our ultraviolet and infrared models as well as the Fermi liquid model. In the latter case, in fact, it can be shown that the V's are defined by convergent series (so that $\mathcal{R}$ is really well defined), if $d = 1$ (see below and also [BG1], §15; [BGPS]). It is therefore possible to proceed by developing the theory at least formally.

This is a somewhat technical aspect that should be ignored until one really tries to perform a mathematically rigorous analysis of some problem, a task that has to be delayed until the conceptual framework which we are trying to set up is fully developed. In practice, in the cases that can be treated rigorously, it turned out that the formal analysis revealed itself as a very important guideline and the difficulties met in elaborating the theory and related to the above boundary conditions (i.e., to keep L finite) have been quite minor ones.

The possibility of defining $\mathcal{R}$, at least via formal expansions, has been exploited very deeply, leading to the possibility of controlling impressive resummations of the (equally divergent but often asymptotic) perturbation series solving formally the problems posed in the previous sections. A *simple* reinterpretation of the old renormalization theory in QFT has been one important consequence of the possibility of defining $\mathcal{R}$ in a precise, yet formal, sense ([FHRW]).

The (6.3) is a map and $1, \mathcal{R}, \mathcal{R}^2, \ldots$ is a family of maps forming a *semigroup* that is usually called the *renormalization group* of our problems. In practice, one will always be interested only in sequences, called *renormalization group flow* or *trajectories*, $\mathcal{R}^{|h|}V$ (with $h \to \pm\infty$ depending on the infrared or ultraviolet character of the problem), which start with a very special V.

Nevertheless, one can try to build one's intuition by imagining to apply $\mathcal{R}$ also to other V's and using concepts of the theory of dynamical systems. One of the first actions taken, when trying to understand the properties of the iterations of a map, is to analyze its fixed points and to study its linearization around them.

The imprecise definition of the domain of $\mathcal{R}$ makes this problem particularly hard when one tries to go beyond formal considerations. In these notes, however, we do not have this ambition and we proceed to reproduce the classical analysis of Wilson.

For instance, to study the scalar field problem, one can begin by remarking that $V = 0$ is a trivial fixed point of the map $\mathcal{R}$ (see (6.3)). The linearization $D\mathcal{R}$ of $\mathcal{R}$ around $V = 0$ is simply

$$D\mathcal{R}\, V(\varphi) = \int P(d\overline{\psi})V(2^{-\delta}\varphi_{2^{-1}}. + \overline{\psi}) , \qquad (6.6)$$

and the same formula holds in the case of the Fermi liquid, with a different meaning of the symbols.

Eigenfunctions of $D\mathcal{R}$ can be easily found. Since $P(d\overline{\psi})$ is a *Gaussian integral*, we expect them to be a kind of generalization of the Hermite polynomials. In fact, the eigenfunctions are related to the Wick monomials which we denote : $\varphi_{x_1} \cdots \varphi_{x_n}$:. The latter are defined recursively by

$$: 1 := 1 , \qquad : \varphi_x := \varphi_x ,$$

$$: \varphi_{x_1} \cdots \varphi_{x_n} := \varphi_{x_1} : \varphi_{x_2} \cdots \varphi_{x_n} : - $$

$$- \sum_{j=2}^{n} C_{x_1 x_j} : \varphi_{x_2} \cdots \varphi_{x_{j-1}} \varphi_{x_{j+1}} \cdots \varphi_{x_n} : , \qquad (6.7)$$

where φ_x is a general Gaussian random field with mean 0 and propagator C_{xy}. It is convenient to regard Wick monomials as defined also when φ is not random (i.e., it is an *external field*), by assigning to φ formally propagator 0 (so that $: \varphi_{x_1} \cdots \varphi_{x_n} := \varphi_{x_1} \cdots \varphi_{x_n}$).

The equations (6.7) clearly extend the recursion relation for the Hermite polynomials and one can, for instance, check that

$$: \varphi_x^n := (2C_{xx})^{n/2} H_n \left(\varphi_x / (2C_{xx})^{1/2} \right) . \tag{6.8}$$

The property that we need here is simply that $\int : \varphi_{x_1} \cdots \varphi_{x_n} :$ $P(d\varphi) \equiv 0$ if $n \neq 0$. This is an easy exercise based on (6.7) and on the Wick rule for Gaussian integrals,

$$\int \varphi_{x_1} \cdots \varphi_{x_n} P(d\varphi) = \sum_{\text{pairings}} \prod_{p \in \text{pairs}} C_p , \tag{6.9}$$

where a *pairing* is a choice of $n/2$ pairs $p = (z, z')$ among $x_1, \ldots, x_n$ with no point common to different pairs (hence n is even, or (6.9) is interpreted as zero).

Another property of the Wick ordered polynomials is their multilinearity. If φ, ψ are independent Gaussian fields, then

$$: (\psi_{x_1} + \varphi_{x_1}) \cdots (\psi_{x_n} + \varphi_{x_n}) := $$
$$= \sum_{\substack{X \subset (x_1, \ldots, x_n) \\ X = (x_{j_1}, \ldots, x_{j_p}) \\ X' = (x_{j_1'}, \ldots, x_{j_{n-p}'})}} : \psi_{x_{j_1}} \cdots \psi_{x_{j_p}} :: \varphi_{x_{j_1'}} \cdots \varphi_{x_{j_{n-p}'}} : , \tag{6.10}$$

valid even if φ or ψ is an external field; then if $f_{x_1 \cdots x_m}(\varphi)$ denotes the monomial $: \varphi_{x_1}^{n_1} \cdots \varphi_{x_p}^{n_p} \partial \varphi_{x_{p+1}}^{n_{p+1}} \cdots \partial \varphi_{x_m}^{n_m} :$, this shows immediately that

$$(DR f_{x_1 \ldots x_m})(\varphi) = 2^{-\delta(n_1 + \cdots + n_p) - (\delta+1)(n_{p+1} + \cdots + n_m)} .$$
$$\cdot f_{2^{-1}x_1 \cdots 2^{-1}x_m}(\varphi) , \tag{6.11}$$

with $\delta = (d - 2 + \gamma)/2$.

Similar definitions and properties hold in the Fermi liquid case. The recursive definition (6.7) is replaced by a new formula in which a sign $\pm$ is inserted in the sum in (6.9) alternating with the parity of the permutation needed to bring the terms of each pair next to each other. Checking this is left to the reader with the hint that everything follows from our definition of Grassmanian integral that *replaces* (6.9) by (3.8) (interpret the terms involving pairings of $++$ fields or of $--$ fields as vanishing because the propagator between such fields has to be considered zero; see (3.8)).

The (6.11) implies immediately that expressions such as

$$F(\varphi) = \int_{(\mathcal{R}^d)^m} dx_1 \dots dx_m W_\beta(x_1, \dots, x_m) \cdot$$

$$\cdot : \varphi_{x_1}^{n_1} \dots \varphi_{x_p}^{n_p} \, \partial\varphi_{x_{p+1}}^{n_{p+1}} \dots \partial\varphi_{x_m}^{n_m} : \tag{6.12}$$

are *eigenfunctions* of $D\mathcal{R}$ if W_β is a homogeneous locally summable function of degree β. In fact the eigenvalues corresponding to them are simply

$$\lambda_F = 2^{\delta_F}, \quad \delta_F = \beta + \sum_{i=1}^{p}(-\delta n_i + d) + \sum_{i=p+1}^{m}(-(\delta+1)n_i + d) , \tag{6.13}$$

which is usually read as saying that the *dimension of the operator F* is the sum of the field dimension ($-\delta$ for each φ and $(-\delta - 1)$ for each $\partial\varphi$) plus the dimension of the volume elements (d each) plus the dimension of the coefficient (β).

One says that F is *relevant* if $\delta_F \geq 0$, $\lambda_F \geq 1$, and *irrelevant* otherwise. F is called an *operator*, without strong philosophical implications.

It is easy to list all the relevant operators even in φ, which are *local*, i.e., which only involve fields evaluated at one point (hence in (6.12) there is only one integration variable). We restrict ourselves only to even monomials, as they are the only ones arising in our a priorieven potentials.

They are the F's, denoted respectively F_2, F_4, $F_{2'}$, F_6, obtained by integrating the following expressions:

$$\begin{array}{ll} : \varphi_x^2 : \; \rightarrow \; \delta_F = 2 - \gamma , & : \varphi_x^4 : \; \rightarrow \; \delta_F = 4 - d - 2\gamma , \\ : (\partial\varphi_x)^2 : \; \rightarrow \; \delta_F = -\gamma , & : \varphi_x^6 : \; \rightarrow \; \delta_F = 6 - 2d - 3\gamma , \end{array} \tag{6.14}$$

whenever the r.h.s. are ≥ 0; or more generally by integrating the : φ_x^{2n} : when their δ_F's, given by $\delta_F = d - n(d - 2 + \gamma)$, are nonnegative.

If we restrict $d \geq 3$ and $\gamma \geq 0$, but $(d, \gamma) \neq (3, 0)$, the relevant terms are among the (6.14). If we let d vary continuously with $d \geq 2$ (the noninteger values have no physical interest), we see that more and more operators become relevant, and if $d = 2$, $\gamma = 0$, *all* the Wick monomials are relevant.

The above list should be enlarged by adding : $\varphi_x \partial \varphi_x$:, with $\delta_F = 1 - \gamma$, or, if $d = 2$ and $\gamma = 0$, also : $\varphi_x^{2n}(\partial \varphi_x)^2$:. But such expressions will turn out never to arise in our theories, or to be reducible to others, so we leave them out.

Much longer would be the list of the nonlocal relevant operators, as in such cases one has the freedom of the choice of β and of the W_β functions.

The above remarks on the eigenfunctions of $D\mathcal{R}$ make it natural to represent the effective potentials V in (6.4), (6.5) by expanding them in Wick monomials rather then in ordinary monomials. Hence we imagine to redefine V_n so that (6.4), (6.5) can be written, respectively,

$$V(\varphi) = \sum_{n,p_1 \cdots p_n} \int V_{n,p_1 \cdots p_n}(x_1, \ldots, x_n) : \Phi_{x_1}^{p_1} \ldots \Phi_{x_n}^{p_n} : d\vec{x} ,$$

$$V(\varphi) = \sum_\alpha \int V_\alpha(x_1, \ldots, x_n) : \Phi_{x_1} \cdots \Phi_{x_n} : e^{ip_F \sum_i \sigma_i \vec{\omega}_i \vec{x}_i} d\vec{x} ,$$

$$\tag{6.15}$$

where $d\vec{x} \equiv dx_1 \ldots dx_n$.

We shall see that one of the main developments of the theory is that, in all our flows $\mathcal{R}^{|h|}V$, the kernels $\overline{V}_n(x_1 \ldots)$ that can arise in (6.15) in dimensionless form will have bounds of the form

$$\left|\overline{V}_n^{(h)}(x_1, \ldots, x_n)\right| \leq C_n e^{-\kappa d(x_1 \cdots x_n)p_0} , \tag{6.16}$$

where p_0 is the momentum scale used in the multiscale decomposition, $\kappa > 0$ is a numerical constant (depending only on the problem considered but not on h, n), and C_n is a constant; $d(x_1 \ldots x_n)$ is the *tree distance* of $x_1, \ldots, x_n$ = length of the shortest tree connecting $x_1, \ldots, x_n$.

The (6.16) will be a bound valid to all orders of a perturbation expansion of $V_n^{(h)}$, when the theory can be worked out only perturbatively (and the dependence of C_n on the order k of perturbation will be rather bad, typically proportional to $k!$).

The (6.16) expresses the key fact that the effective potentials (dimensionless) keep a fixed range, finite, uniformly in the scale parameters to all orders of perturbation theory. Hence the myriad of relevant nonlocal operators are not represented in the expansions (6.14).

In the following chapter, we will examine the implications of the above analysis. We shall conclude this chapter by mentioning that in the case of Fermi liquids a similar analysis can be performed and one finds that

the only local operators that can arise in the theory's effective potentials
are the following, if $d = 1$:

$$F_1 = \int \; : \psi^+_{x+1}\psi^+_{x-1}\psi^-_{x+1}\psi^-_{x-1} : \, dx \; ,$$

$$F_2 = \int \; : \psi^+_{x\bar{\omega}_1}\psi^-_{x\bar{\omega}_2} : e^{i(\bar{\omega}_1-\bar{\omega}_2)p_F\bar{x}} \, dx \; ,$$

$$F_3 = \int \bar{\omega}_2 \cdot : \psi^+_{x\bar{\omega}_1}\vec{D}_{\bar{\omega}_2}\psi^-_{x\bar{\omega}_2} : e^{i(\bar{\omega}_1-\bar{\omega}_2)p_F\bar{x}} \, dx \; ,$$

$$F_4 = \int \; : \psi^+_{x\bar{\omega}_1}\partial_t\psi^-_{x\bar{\omega}_2} : e^{i(\bar{\omega}_1-\bar{\omega}_2)p_F\bar{x}} \, dx \; .$$

(6.17)

Note that there is only one local term of fourth order in the field because
of the Fermi statistics and the fact that $\omega = \pm 1$ and that our fermions
are spinless.

The operators (6.17), if the $\bar{\omega}$'s are fixed so that $\bar{\omega}_1 - \bar{\omega}_2 = 0$, are
eigenfunctions of $D\mathcal{R}$ with eigenvalues $\lambda_F = 2^{\delta_F}$ with

$$\delta_{F_1} = 0, \quad \delta_{F_2} = 1, \quad \delta_{F_3} = \delta_{F_4} = 0 \; , \quad (6.18)$$

with only one relevant operator.

The case of the Bose gas is similar and is discussed in chapters 13
and 14.

Chapter 7
Asymptotic Freedom:
Upper Critical Dimension

The implications of the analysis of the previous section can be easily appreciated in the ultraviolet problem or in the infrared problem in the simple cases with $\gamma > 0$, which we assume below (unless explicitly stated that the case considered is $\gamma = 0$).

Assume that the couplings (*bare couplings*) defining the initial V are small. Then we want to see if the $V^{(h)}$ stay consistently small as h evolves to ∞ ($-\infty$ in the infrared cases, $+\infty$ in the ultraviolet), at least if one uses the linearized approximation for $\mathcal{R}$, (6.6).

In this approximation we simply have to write the initial V in dimensionless form $\overline{V}$ (see chapter 6), and express $\overline{V}$ as a superposition of the eigenvectors of $D\mathcal{R}$.

Taking into account that V has the form (4.2), and calling, as in chapter 6, $\overline{\psi}$ the dimensionless field variables, we find in the scalar cases

$$\overline{V}(\overline{\psi}) = \int \left(\overline{\lambda} : \overline{\psi}_x^4 : + \overline{\mu} : \overline{\psi}_x^2 : \right) dx \tag{7.1}$$

(see (2.9), (2.12) and recall that $: \varphi_x^4 := \varphi_x^4 - 6C_{xx}\varphi_x^2 + 3C_{xx}^2, \; : \varphi_x^2 := \varphi_x^2 - C_{xx}$), with

$$\begin{cases} \overline{\lambda} = L_0\beta^{-2} \\ \overline{\mu} = 6\overline{\lambda}C_{xx} + M_0(\beta)\beta^{-1} \end{cases} \qquad \begin{cases} \overline{\lambda} = \lambda_N 2^{-(4-d-2\gamma)N} \\ \overline{\mu} = (\mu_N + 6\lambda_N C_{xx}^{(\leq N)})2^{-(2-\gamma)N} \end{cases} \tag{7.2}$$

the left relations pertaining to the infrared problem and the right to the ultraviolet.

In the linearized flow, the effective potential on scale h can be written as $\overline{V}^{(h)} = D\mathcal{R}^{|h|}\overline{V}$ with the scale label $h = 0, -1, -2, \ldots, -\infty$ in the infrared case, while it is $\overline{V}^{(N-h)} = D\mathcal{R}^h\overline{V}$, $h = N, N-1, \ldots, 0$, in the ultraviolet case. In fact $\overline{V}^{(h)}$ have, for the appropriate values of h, still the form (7.1) with $\overline{\lambda}$, $\overline{\mu}$ replaced by $\overline{\lambda}_h$, $\overline{\mu}_h$ verifying (see (6.14))

$$\overline{\lambda}_{h-1} = 2^{4-d-2\gamma}\overline{\lambda}_h \;, \qquad\qquad \overline{\mu}_{h-1} = 2^{2-\gamma}\overline{\mu}_h \;. \tag{7.3}$$

Hence we see that our hypothesis that $\overline{\lambda}_h$ and $\overline{\mu}_h$ stay small for all h's of interest has a different meaning in the two cases.

In the infrared problem the latter hypothesis leads to the conditions $\overline{\mu}_h \equiv 0$ and $4-d-2\gamma < 0$. Expressing this in terms of the range exponent of the spin potential $J(x) \sim 1/|x|^\alpha$, recalling (2.6) with $\gamma = d+2-\alpha$, we see that this means

$$\alpha < \frac{3}{2}d \qquad (7.4)$$

(and our assumption $2 > \gamma > 0$ means $d < \alpha < d+2$); hence (7.4) is a condition that the potential should have "very long range." Note that if $d \geq 4$, "very long range" is *any* power law potential (as $4 - d - 2\gamma < 0$ for $\gamma > 0$).

Of course, in the linear approximation the μ_h grow exponentially, if $\mu_0 \equiv \overline{\mu} \neq 0$, and we have to expect that the condition $\overline{\mu} = 0$ is modified by the nonlinear terms into a condition like $\overline{\mu} = \overline{\mu}(\overline{\lambda})$, i.e., that there should be *only one* special choice of $\overline{\mu}$ for which it will be possible that the effective potentials, in dimensionless form, stay uniformly bounded and small.

And if $\overline{\mu}$ is so chosen (i.e., $\overline{\mu} = 0$ in the linear approximation, with (7.4) verified), then $\lambda_h \to 0$, $\mu_h \to 0$ as $h \to -\infty$ and we see that for such value of $\overline{\mu}$ the correction to the $S^T(x_1, x_2)$, i.e., to the pair correlation function, on large scales vanishes ($\overline{\lambda}_\infty, \overline{\mu}_\infty \equiv 0$) and the pair correlation function decays as the free propagator, rather than like the potential, signaling that we are at the critical point (see comment following (2.16)). In the linear approximation, the (7.2), (2.9) and the condition $\overline{\mu} = 0$ give us even the value of β_c as the solution of

$$6L_0\beta^{-2}C_{xx} + (-R_0 + \beta r)\beta^{-1} = 0 , \qquad (7.5)$$

which has a solution if $r < R_0^2/24L_oC_{xx}$.

The above conclusion is quite exciting, as it provides a scheme to calculate the critical point whose existence and nature are sensitive to the space dimension and to the range of the potential. We realize also that we have found that the models considered have a *trivial* or *free* or *mean field* critical point, if the interaction has long enough range or if $d > 4$ (always taking $\gamma > 0$, i.e., a polynomial decay of the pair potential).

The cases with $\alpha - d - 2\gamma = 0$ and $\gamma \geq 0$ will be discussed later: note that the property that $\overline{\lambda}_h, \overline{\mu}_h$ stay small for all h's is not decidable to first order because the expansion rate of λ will be exactly 1 (*marginality* of λ).

Leaving aside, for the moment, the cases $d + 2 - \gamma = 0$ or $\gamma = 0$, note that the above discussion gives no information about the case $\overline{\mu} \neq \overline{\mu}(\overline{\lambda})$ (or $\beta \neq \beta_c$). In fact $\overline{\mu}_h$ is in this case growing so that, after a few iterations of $\mathcal{R}$, there cannot be any ground to proceed as if the higher order corrections to the approximation $\mathcal{R} \sim D\mathcal{R}$ were negligible.

Let us now look at the implications of the above analysis in the case of the ultraviolet problem. In this case there is no γ. However we are interested in a different question. Namely, $\overline{\lambda}$, $\overline{\mu}$ are free (as such are λ_N, μ_N) and we must show that they can be so chosen that after N iterations of the renormalization map we built an effective potential (small, of course, to be consistent with our linear approximation) that is nontrivial.

In our approximation, this simply means that λ_0, μ_0 have to be finite (small) and $\lambda_0 \neq 0$ (to make sure that the effective potential $V^{(0)}$ is not a quadratic form). Manifestly if $d = 2, 3$ (i.e., $d < 4$), this is possible, and in many ways. Namely, one just fixes $\overline{\lambda}_0, \overline{\mu}_0$ and determines $\overline{\lambda}, \overline{\mu}$ as

$$\overline{\lambda}_N \equiv \overline{\lambda} = \overline{\lambda}_0 2^{-(4-d)N} \ , \qquad \overline{\mu}_N \equiv \overline{\mu} = \overline{\mu}_0 2^{-2N} \ , \qquad (7.6)$$

which is perfectly acceptable, as this means that $\overline{\lambda}_h$, $\overline{\mu}_h$ are small for all values of $h = 0, 1, \ldots, N$ and in fact very small for most h's.

The evolution of the *running coupling constants*, $\overline{\lambda}_h$, $\overline{\mu}_h$, in the above cases is thus such that one can find initial data for $\overline{V}$ that generate a family of running coupling constants which approaches zero as their scale labels h go to ∞ ($-\infty$ in the infrared case or $+\infty$ in the ultraviolet case). At the same time, they have nonzero values on scales of order 1 ($|h| \sim 1$), and the constants' value relative to the nonquadratic relevant operators is not zero (an important condition, as it excludes triviality).

One says that the above infrared problem under condition (7.4) and the ultraviolet problem with $d < 4$ are *asymptotically free*.

More generally one says that a problem is *asymptotically free* if

1. one can find a way to go systematically beyond the linear approximation $D\mathcal{R}$ to $\mathcal{R}$ by successive approximations, and
2. the first approximation to $\mathcal{R}$ for which the nonquadratic relevant operators evolve nontrivially starting from a suitably chosen small initial $\overline{V}$ has the property that the corresponding dimensionless couplings approach zero for large values of the scale labels, although they have a nonzero (small) value at scales of order 1; at the same time the other relevant couplings remain small and approach zero, or at least

stay bounded and small over the whole range of scales.

Most cases of physical interest have, however, the feature that the asymptotic freedom property does not hold or its validity cannot be decided by simply looking at the linear approximation of $\mathcal{R}$.

The reason for this is that there exist *marginal operators* that are not quadratic in the fields and which, in the linear approximation, have running couplings that stay constant (i.e., have $\delta_F = 0$). We have already met the most prominent examples:

1. the infrared problem with $0 < \gamma < 2$ and $4 - d - 2\gamma = 0$,
2. the infrared problem with $\gamma = 0$, $d = 4$, and
3. the ultraviolet problem with $d = 4$,

which are listed in order of increasing difficulty. Later we shall discuss the Bose condensation problem.

Case 1 is asymptotically free to second order. Case 2 is also asymptotically free but only the running constants associated with the non-quadratic relevant operators approach 0 as $h \rightarrow -\infty$. Case 3 is very famous, as it is the simplest example of a nonasymptotically free model whose asymptotic freedom cannot be decided in the linear approximation (as it is the case in the infrared problems with $4 - d - 2\gamma > 0$ or in the $d > 4$ ultraviolet problems).

4. The $d = 1$ (spinless) Fermi liquid problem is a fourth problem for which the asymptotic freedom *cannot* be decided from the linear approximation (see (6.18)). And it is particularly interesting because its asymptotic freedom cannot be decided in any finite order approximation. As we shall discuss, eventually, it will turn out that it is *not* asymptotically free.
5. The Bose condensation problem, which we discuss in chapter 15, represents a remarkable instance in which in some appropriate sense asymptotic freedom holds.

Chapter 8
Beyond the Linear Approximations: The Beta Function and Perturbation Theory

The beta function is a powerful tool to investigate properties of improperly convergent series expansions and to turn perturbation theory in statistical mechanics or quantum field theory into a (sometimes) useful algorithm.

In fact, until recently it was not clear how to turn the enormous amount of work performed in renormalization theory of quantum fields into a constructive algorithm. This was similarly true for the theory of perturbations in classical mechanics before the development of the *KAM theory* out of the work of Kolmogorov (which, as already mentioned, is also an *ante litteram* instance of an application of renormalization group ideas).

The beta function arises when one tries to find the corrections to the linear approximation to the renormalization map $\mathcal{R}$. It is particularly useful in the *linearly undecidable* cases listed at the end of chapter 7.

The idea is that the (dimensionless) potential on scale h, $\bar{V}^{(h)}$, can be written as

$$\bar{V}^{(h)} = \mathcal{L}\bar{V}^{(h)} + (1 - \mathcal{L})\bar{V}^{(h)} , \tag{8.1}$$

where $\mathcal{L}$ is a projection operator extracting out of $\bar{V}^{(h)}$ its *relevant* components. So $\mathcal{L}$ is a linear operator with often finite dimensional range, coinciding with the linear span of the relevant operators (which, in the four linearly undecidable cases of chapter 7, are finitely many).

At this point one can ask how is $\mathcal{L}$ determined if one knows nothing. In the early days of the renormalization group methods, it was often stated that basically there is no way to tell a prioriwhich is the renormalization transformation to use: the good one is that which works! [1]

---

[1] To no one's surprise, loosely speaking one can say that the renormalization group *cannot* be an easy way to solve a problem: finding a renormalization group transformation that solves a problem is always preceded (to our knowledge) by an understanding of the basic features of the problem, at least at an intuitive or phenomenological level.

Thus the first thing one has to realize is that the formalism is absolutely general, and one has as much freedom as one can imagine in defining the operator $\mathcal{L}$. Therefore it is important to understand the ideas *without* really fixing $\mathcal{L}$.

It turns out, however, that after a while one realizes that $\mathcal{L}$ is always the same: and the real difficulty is to find an "appropriate" functional integral formulation of the problem, at least for the problems that can be formulated as functional integrals.[2] The "old" ambiguity in the choice of the renormalization transformation goes into that of finding the appropriate functional integral formulation (this is really well illustrated by the Bose condensation problem).

Hence we shall illustrate the general formalism with a special choice of $\mathcal{L}$, i.e., the one that will turn out to be the *correct* one (essentially unique up to trivialities, as far as we know). But no real use is made of the special form of $\mathcal{L}$, unless when we discuss bounds (which can only be found in a useful form if the choice of $\mathcal{L}$ is the indicated one).

The reader should read what follows in this spirit. The fact that the choice of $\mathcal{L}$ is correct is expressed by the bound (8.28) for the cases discussed in this section. Its proof is a nontrivial (although not really hard) task, essentially identical to a proof of renormalizability of a scalar field theory: *we shall not give the proof of (8.28) in the cases at hand.* The validity of (8.28) (i.e., the scale independence of the beta function) to second order of the perturbation expansion is often considered by most physicists as good enough for a formal theory.[3]

The operator $\mathcal{L}$ has to be devised, if possible, so that the *irrelevant part of* $V^{(h)}$, $(1 - \mathcal{L})V^{(h)}$, can be expressed in terms of the values of the coefficients $\vec{v}_{h'}$ of the relevant operators in $\mathcal{L}V^{(h')}$, $h' > h$; if this is possible, the problem of the analysis of the sequence[4] $\mathcal{R}^n \bar{V}$, $n \geq 0$

---

[2] Not all the problems listed in the introduction are reducible to a functional integration, although in the future they might be formulated as such. For instance, recently, and quite surprisingly, the problem of the determination of the invariant tori in the KAM theory was formulated as a functional integral (see [G3]).

[3] One should add that they are aware that this makes sense only if a scale independent bound like (8.28) holds, and sometimes it is declared to be obviously true, "by the renormalizability theorems."

[4] It is perhaps worth stressing that in all cases $n \geq 0$. The transformation $\mathcal{R}^{-1}$ is never studied. The map $\mathcal{R}$ should always be regarded a noninvertible one in the same sense in which the evolution governed by the heat equation is not invertible: the forward evolution is very nice, while the backward evolution is, even when it

becomes that of understanding the properties of the sequence $\vec{v}_h$ of the *running coupling constants*.

To define $\mathcal{L}$, in the scalar case, we imagine $V^{(h)}$ expressed as in (6.4) and define $\mathcal{L}$ by linearity, prescribing its action on the Wick monomials. Considering only the cases $4 - d - 2\gamma \geq 0$, $\gamma \geq 0$, $d \geq 3$ in the infrared problem and $d = 3, 4$ in the ultraviolet, we define

$$\text{a.} \quad \mathcal{L} : \varphi_{x_1} \varphi_{x_2} \varphi_{x_3} \varphi_{x_4} := \frac{1}{4} \sum_{j=1}^{4} : \varphi_{x_j}^4 : \tag{8.2}$$

$$\text{b.} \quad \mathcal{L} : \varphi_{x_1} \varphi_{x_2} := \frac{1}{2} \Big(: \varphi_{x_1}^2 : + (x_2 - x_1) : \varphi_{x_1} \partial \varphi_{x_1} : + $$
$$+ \frac{1}{2}(x_2 - x_1)^2 : \varphi_{x_1} \partial^2 \varphi_{x_1} : \Big) + (1 \longleftrightarrow 2)$$

if $\gamma = 0$ and, more simply, (b) can be defined to be

$$\text{b'.} \quad \mathcal{L} : \varphi_{x_1} \varphi_{x_2} : = \frac{1}{2} \big(: \varphi_{x_1}^2 : + (x_2 - x_1) : \varphi_{x_1} \partial \varphi_{x_1} : \big) + $$
$$+ (1 \leftrightarrow 2) \tag{8.3}$$

when $\gamma > 0$, $\gamma \leq 1$ or

$$\text{b''.} \quad \mathcal{L} : \varphi_{x_1} \varphi_{x_2} := \frac{1}{2} \sum_{j=1}^{2} : \varphi_{x_j}^2 : \tag{8.4}$$

when $\gamma > 1$.

Similarly in the case of Fermi liquids (and $d = 1$), we define

$$\mathcal{L} : \psi_{x_1 \vec{\omega}_1}^+ \psi_{x_2 \vec{\omega}_2}^+ \psi_{x_3 \vec{\omega}_3}^- \psi_{x_4 \vec{\omega}_4}^- := \frac{1}{2} \sum_{j=1}^{2} : \psi_{x_j \vec{\omega}_1}^+ \psi_{x_j \vec{\omega}_2}^+ \psi_{x_j \vec{\omega}_3}^- \psi_{x_j \vec{\omega}_4}^- : \,,$$
$$\mathcal{L} : \psi_{x_1 \vec{\omega}_1}^+ \psi_{x_2 \vec{\omega}_2}^- : = $$
$$=: \psi_{x_1 \vec{\omega}_1}^+ \psi_{x_2 \vec{\omega}_2}^- : + (x_2 - x_1) : \psi_{x_1 \vec{\omega}_1}^+ D_{\vec{\omega}_2} \psi_{x_2 \vec{\omega}_2}^- : \,, \tag{8.5}$$

where $D_{\vec{\omega}} = (\partial_t, \partial_{\vec{x}} - i\vec{\omega}\Delta_{\vec{x}}/2p_F)$ is a convenient differential operator; we could use $\partial = (\partial_t, \partial_{\vec{x}})$ instead of $D_{\vec{\omega}}$, but the latter is more natural and

could be defined, in many ways singular (and of little interest for our purposes). The fact that the *same* transformation is interesting for widely different problems, like the infrared problem of the critical point and the ultraviolet problems of scalar field theory, is certainly a unifying feature justifying the interest that the renormalization group methods have elicited in theoretical physics.

its use simplifies the calculations, as one can realize by trying to do some of them. The $D_{\bar{\omega}}$ operator is a kind of *covariant derivative* in the sense that its consistent use in the calculations keeps (more) manifest the basic property that the quasi particles have been introduced as a technical device to study effective potentials *which only depend on particle fields*; this implies that the dependence on the $\bar{\omega}$-variables is somewhat special as it always has to be possible to perform the $\bar{\omega}$-integrals to reexpress the $V^{(h)}$ as functions of particle fields.

We define $\mathcal{L}$ to be zero on the monomials of degree higher than 4.

It is easy to see that, with the notation defined in the lines above (6.14),

$$\begin{cases} \mathcal{L}V^{(h)}(\phi) = \lambda_h F_1(\phi) + \mu_h F_2(\phi) + \alpha_h F_3(\phi) & \text{if } \gamma = 0 \\ \mathcal{L}V^{(h)}(\phi) = \lambda_h F_1(\phi) + \mu_h F_2(\phi) & \text{if } \gamma > 0 , \end{cases} \tag{8.6}$$

where no operator with : $\varphi_x \partial \varphi_x$: appears, as such terms can be integrated by parts $(\varphi_x \partial \varphi_x = 2^{-1} \partial \varphi_x^2)$ and disappear; and in the Fermi liquid case $(d = 1)$,

$$\mathcal{L}V^{(h)} = \lambda_h F_1 + \sum_{\omega\omega'}(\nu_h F_2 + \alpha_h F_3 + \zeta_h F_4) . \tag{8.7}$$

Here, we have used (6.7), (6.17).

In the scalar cases we see that $\mathcal{L}$, as defined above, indeed projects on the space of the relevant operators: *with the exception* of the case $d = 3, \gamma = 0$, and this is a real problem in the infrared case because the sixth-order operators : φ_x^6 :, as soon as they are created by the nonlinear terms, will remain forever and will give important contributions to the further evolution of the effective potential. This is less serious in the ultraviolet problem because there we have already seen that we have to take $\bar{\lambda}_N$ and $\bar{\mu}_N$ extremely small: this implies that the sixth-order relevant terms generated are very small and do not contribute much to the effective potentials.

It is, however, a very serious difficulty in the infrared problem: therefore we exclude the infrared case $d = 3, \gamma = 0$, which is clearly more difficult than the others. We will come back to it (see chapter 12), and we shall give up playing the game of trying to modify the definition of $\mathcal{L}$ to fix the just-mentioned difficulty (the reader should try this as an exercise). In fact, it will appear that such direction is not at all a promising one.

In the Fermi liquid case we see that $\mathcal{L}$ projects on operators containing components on F_2, F_3, F_4 that are sums of eigenvectors of $D\mathcal{R}$

(corresponding to $\bar{\omega}_1 = \bar{\omega}_2$ in (6.17)) and of operators that are not eigenvectors of $D\mathcal{R}$ (corresponding to $\bar{\omega}_1 = -\bar{\omega}_2$). It is convenient, however, not to modify the definition of $\mathcal{L}$ to impose that it strictly projects over the relevant eigenvectors: in this way, in fact, one keeps more easily track of a kind of *gauge* invariance property. The latter property is a deep consequence of the fact that we know *a priori* that all our effective potentials can be expressed in terms of the particle fields, since the original potential had this property. It also implies that the running couplings are in fact independent of ω, ω', as in (8.7).

The important point is that $\mathcal{L}$ extracts out of $V^{(h)}$ the relevant part, and the fact that it also extracts a piece of what we would like to consider irrelevant does not affect the discussion. In general, one may expect to be free, at least to some extent, to decide that some irrelevant operators are treated as relevant and that this should not really affect the analysis.

Having defined the operators $\mathcal{L}$, we come to the following problems:
1. Find an expansion of $(1 - \mathcal{L})V^{(h)}$ in powers of $\vec{v}_{h'}$, $h' > h$.
2. Find the relation between $\vec{v}_h$ and $\vec{v}_{h-1}$.

If (1) can be solved, the whole problem will have been reduced to that of studying the sequence of the running couplings $\vec{v}_h$.

Basically the only known method to study (1), (2) is perturbation theory.

Let $\bar{V}$ be the initial dimensionless potential. We begin by decomposing it into relevant and irrelevant parts, and this is accomplished by applying $\mathcal{L}$ and $1 - \mathcal{L}$.

In the scalar cases, $\bar{V} \equiv \mathcal{L}\bar{V}$, obviously. This is not the case in the Fermi liquid problem. In this case, in fact, we recall that

$$
\bar{V} = \int \lambda v(\vec{x} - \vec{y})\delta(x_0 - y_0) : \psi_x^+ \psi_x^- \psi_y^+ \psi_y^- : dx\,dy +
$$
$$
+ \int \nu : \psi_x^+ \psi_x^- : dx + \int \alpha : \psi_x^+ \frac{-\Delta - p_F^2}{2m} \psi_x^- : dx \ , \tag{8.8}
$$

and we see that the quartic part of $\overline{V}$ is nonlocal.

The (8.8) is slightly more general than the one considered so far (see the third of (4.2) which had $\nu = \alpha = 0$). The reason for this more general starting point is that the two extra parameters can be roughly interpreted as variations of the chemical potential ($\mu \to \mu + \nu$) and of the mass of the particles $1/m \to (1 + \alpha)/m$. Therefore, introducing two such parameters allows us to think that they can be conveniently fixed so that the long-range behavior ("physically observable") of the $S(x)$ function corresponds indeed to a singularity in k-space at $|k| = p_F$, $k_0 = 0$

describing particles of physical mass m (in a sense to be understood). In others words, adding two parameters ν, α has the purpose of permitting us to fix a priorithe physical values p_F and m of the Fermi momentum and of the mass (foreseeing that they could change because of the interaction).

The Wick ordering in (8.8) can be put in or taken out at will at the price of suitably changing the quadratic term (possibly making it nonlocal).

Using (5.13) we see that $\tilde{V}$, (8.8), becomes, in terms of the quasiparticles fields,

$$\sum_{\bar{\omega}_1 \cdots \bar{\omega}_4} \int \lambda v(\vec{x} - \vec{y}) \delta(x_0 - y_0) e^{i(\bar{\omega}_1 - \bar{\omega}_3)p_F \vec{x} + (\bar{\omega}_2 - \bar{\omega}_4)p_F \vec{y}}.$$

$$\cdot : \psi^+_{x\bar{\omega}_1} \psi^-_{x\bar{\omega}_3} \psi^+_{y\bar{\omega}_2} \psi^-_{y\bar{\omega}_4} : dx \, dy + \sum_{\bar{\omega}_1 \bar{\omega}_2} \int \left(\nu : \psi^+_{x\bar{\omega}_1} \psi^-_{x\bar{\omega}_2} : + \right. \tag{8.9}$$

$$\left. + \alpha : \psi^+_{x\bar{\omega}_1} (-i\bar{\partial}_2) D_{\bar{\omega}_2} \psi^-_{x\bar{\omega}_2} : \right) e^{i(\bar{\omega}_1 - \bar{\omega}_2)p_F \vec{x}} dx \, ,$$

thus, applying (8.5), we find, recalling (6.17) and after a short calculation,

$$\mathcal{L}\bar{V} = \lambda_0 F_1 + \sum_{\bar{\omega}\bar{\omega}'} [\nu_0 F_2 + \alpha_0 F_3 + \zeta_0 F_4] \, , \tag{8.10}$$

$$\lambda_0 = 2\lambda[\hat{v}(0) - \hat{v}(2p_F)], \qquad \nu_0 = \nu, \qquad \alpha_0 = \alpha, \qquad \zeta_0 = 0 \, ,$$

while

$$(1 - \mathcal{L})\bar{V} = \sum_{\bar{\omega}_1 \cdots \bar{\omega}_4} \int \lambda v(\vec{x} - \vec{y}) \delta(x_0 - y_0) \cdot$$

$$\cdot \, e^{i(\bar{\omega}_1 - \bar{\omega}_3)p_F \vec{x} + i(\bar{\omega}_2 - \bar{\omega}_4)p_F \vec{y}} \, . \tag{8.11}$$

$$\cdot \left[: \psi^+_{x\bar{\omega}_1} \psi^-_{x\bar{\omega}_2} (\psi^+_{y\bar{\omega}_3} - \psi^+_{x\bar{\omega}_3}) \psi^-_{y\bar{\omega}_3} : + : \psi^+_{x\bar{\omega}_1} \psi^-_{x\bar{\omega}_2} \psi^+_{x\bar{\omega}_3} (\psi^+_{y\bar{\omega}_3} - \psi^-_{x\bar{\omega}_3}) : \right] \, .$$

The role of the initial irrelevant part (8.11) in the following analysis is a minor one if one forgets about the *gauge symmetry*, i.e., the a priori-obvious fact that $V^{(h)}$ has to be expressible in terms of particles fields. The latter is the reason we do not make here the further *simplification* of simply discarding it.

To proceed, it is better to develop the formalism first in the scalar cases where in $\bar{V}$ no *irrelevant* terms are present (calling "irrelevant" also $(1 - \mathcal{L})\bar{V}$, a name which will be justified by the coming analysis).[5]

---

[5] It wil become clear that "irrelevant" does not at all mean negligible; in fact, the physically interesting quantities are just expressed in terms of the truncated Schwinger functions which are simply related to the irrelevant operators.

We set up a recursive scheme to evaluate $\mathcal{R}$ as follows. Let $X_1, \ldots, X_p$ be p random variables and denote by $\mathcal{E}$ the integration over their (joint) distribution. Then we introduce the *truncated expectations* $\mathcal{E}^T(X_1, \ldots, X_p)$, sometimes called *cumulants* or *connected expectations*, as

$$\mathcal{E}^T(X_1, \ldots, X_p) = \frac{\partial^p}{\partial \varepsilon_1 \ldots \partial \varepsilon_p} \log \mathcal{E} \left(e^{\sum_i \varepsilon_i X_i} \right)\Big|_{\varepsilon_i = 0} , \qquad (8.12)$$

which certainly makes sense if $e^{c|X_i|}$ are $\mathcal{E}$-integrable for all $c > 0$; but it is clear that (8.12) is a finite combination of products of expectations (of products of some) of the X_i's; hence in fact (8.12) makes sense under the *much weaker* hypothesis that $|X_i|^q$ is $\mathcal{E}$-integrable for all $q > 0$.

Then, formally,

$$-\mathcal{R}\bar{V}(\bar{\varphi}) = \sum_{p=1}^{\infty} \frac{1}{p!} \mathcal{E}^T \overbrace{\left(-\bar{V}(2^{-\delta}\bar{\varphi}_{2-1.} + \cdot), \ldots, -\bar{V}(2^{-\delta}\bar{\varphi}_{2-1.} + \cdot)\right)} , $$

$$(8.13)$$

where the number of arguments under the curly bracket are p.

Note that (8.13) is nothing else (see (8.12)) but the development of the expression $\log \mathcal{E}(e^{-t\bar{V}(2^{-\delta}\varphi_{2-1.} + \cdot)})$ developed in a formal Taylor series and evaluated at $t = 1$.

To proceed, it is more convenient to work with the dimensional effective potentials making them dimensionless only when the results have to be expressed. Thus calling V the dimensional potential corresponding to $\bar{V}$, we rewrite (8.13) as

$$-\mathcal{R}V(\varphi) = \sum_{p=1}^{\infty} \frac{1}{p!} \mathcal{E}^T(-V(\varphi + \cdot), \ldots, -V(\varphi + \cdot)) . \qquad (8.14)$$

To fix the ideas, let us investigate the ultraviolet problem. Then $V \equiv V^{(N)}$, $\bar{R}V \equiv V^{(N-1)}$, $\mathcal{E} \equiv \mathcal{E}_N$ denotes the integration with respect to $\psi^{(N)}$ and (8.14) is

$$-V^{(N-1)}(\varphi) = \sum_{p=1}^{\infty} \frac{1}{p!} \mathcal{E}_N^T(-V^{(N)}(\varphi + \cdot), \ldots, -V^{(N)}(\varphi + \cdot))$$

$$(8.15)$$

$$\equiv \sum_p V_p^{(N-1)}(\varphi) .$$

We iterate the above relation. This rather involved operation can be visualized by using the following graphical representation:

$$-V^{(N)} \; \rightarrow \qquad \underset{N}{\bullet\!\!-\!\!\!-\!\!\!-\!\!\!-} \tag{8.16}$$

$$-V^{(N-1)} = \sum_{p=1}^{\infty} \frac{1}{p!} \mathcal{E}_N^T \overbrace{(-V^{(N)}, \ldots, -V^{(N)})} \; \rightarrow \; \sum_{p=1}^{\infty} \underset{N-1 \quad N}{\bullet\!\!-\!\!\!-\!\!\!\!<\!\!\!} \tag{8.17}$$

where the number of terms under the curly bracket is p and the number of endlines in the second graph is also p.

Thus (8.15) can be represented as

$$-V^{(N-1)} = \underset{N-1 \quad N}{\bullet\!\!-\!\!\!-\!\!\!-\!\!\!\bullet} \;+\; \underset{N-1 \quad N}{\bullet\!\!-\!\!\!\!<} \;+\; \underset{N-1 \quad N}{\bullet\!\!-\!\!\!\!<\!\!\!} \;+\ldots = \tag{8.17}$$

$$= \underset{N-1}{\bullet\!\!-\!\!\!-\!\!\!\bigcirc}$$

so that, with the same notation,

$$-V^{(N-2)} = \underset{N-2 \quad N-1}{\bullet\!\!-\!\!\!-\!\!\!-\!\!\!\bullet\!\!-\!\!\!-\!\!\!\bigcirc} \;+\; \underset{N-2 \quad N-1}{\bullet\!\!-\!\!\!-\!\!\!-\!\!\!\bullet\!\!\!<} \;+\ldots \tag{8.18}$$

and then using (8.17) and taking into account the multilinearity of the $\mathcal{E}^T$ (see (8.12)) we can write

$$-V^{(N-2)} = \underset{N-2\ \ N-1\ \ \ N}{\bullet\!\!-\!\!\!-\!\!\bullet\!\!-\!\!\!-\!\!\bullet} \quad + \quad \underset{N-2\ \ \ N-1\ \ \ \ N}{\bullet\!\!-\!\!\!-\!\!\bullet\!\!-\!\!\!-\!\!\bullet{<}} \quad +..+$$

$$(8.19)$$

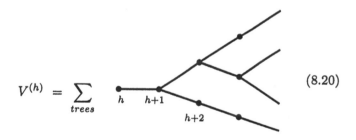

Therefore,

$$V^{(h)} = \sum_{trees}$$

$$(8.20)$$

and we see that $V^{(h)}$ can be represented in terms of the planar trees with an arbitrary number of vertices arranged at points of abscissa $h, h+1, \ldots, N$ at each of which there is a bifurcation into $1, 2, 3, \ldots$ branches; the branches go all the way up to the highest scale index. The vertices symbolize truncated expectations of whatever symbolizes every line emerging from the vertex (we *read* the tree from left to right). Thus, for instance, the tree

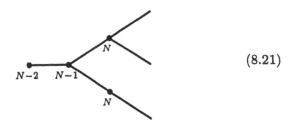

(8.21)

means

$$\frac{1}{2!}\mathcal{E}^T_{N-1}\left(\frac{1}{2!}\mathcal{E}^T_N(-V^{(N)}, -V^{(N)}), \mathcal{E}^T_N(-V^{(N)})\right),\qquad (8.22)$$

where $V^{(N)} = V^{(N)}(\varphi + \psi^{(N-1)} + \psi^{(N)})$ and $\mathcal{E}_N, \mathcal{E}_{N-1}$ denote the integrations with respect to the distributions of $\psi^{(N)}$ and $\psi^{(N-1)}$.

Clearly the above procedure produces a power-series expansion of $V^{(h)}$ in powers of the couplings $\bar{v}_N$; it is the usual *nonrenormalized* expansion of the effective potential $V^{(h)}$ and, if we set, for $h = -1$, $V^{(-1)}(\varphi) \equiv -V_{\text{eff}}(\varphi)$, it gives the usual perturbation expansion of QFT, with its usual problems, i.e., divergences as $N \to \infty$.

But of course our program is to build a different series, namely, a series in the running couplings. The above algorithm has the great advantage of being designed exactly to construct such expansion from the previous one, essentially at no extra work.

The procedure is the following: construct $V^{(N-1)}$ as in (8.17). Apply $\mathcal{L}$ to $V^{(N-1)}$ and $(1-\mathcal{L})$ to each of the terms in (8.17) composing $V^{(N-1)}$, and denote

$$-\mathcal{L}V^{(N-1)} = \underset{N-1}{\bullet\!-\!-\!-\!-}$$

$$-(1-\mathcal{L})V^{(N-1)} = \underset{N-1}{\bullet}\overset{\mathcal{R}}{-\!\!-}\underset{N}{\bullet} \quad + \quad \underset{N-1}{\bullet}\overset{\mathcal{R}}{\underset{N}{-\!\!\!<}} \quad + \quad (8.23)$$

$$+ \quad \underset{N-1}{\bullet}\overset{\mathcal{R}}{\underset{N}{-\!\!\!<}} \quad +\dots$$

Noting that the first term in (8.17) just represents the action of the linearized renormalization map, hence the action of $(1-\mathcal{L})$ on it gives 0, we can rewrite (8.17) as:

$$-V^{(N-1)} = \underset{N-1}{\bullet\!\!-\!\!-\!\!\bullet} \quad + \quad \overset{\mathcal{R}}{\underset{N-1\quad N}{\bullet\!\!-\!\!\bullet\!\!<}} \quad + \quad \overset{\mathcal{R}}{\underset{N-1\quad N}{\bullet\!\!-\!\!\bullet\!\!<}} \quad +\dots \ (8.24)$$

where the first term represents $\mathcal{L}V^{(N-1)}$; hence it is determined by the running coupling $\vec{v}_{N-1}$ and the others add up to $(1-\mathcal{L})V^{(N-1)}$.

We now iterate the procedure; clearly the result is

$$\mathcal{L}V^{(h)} = \underset{h\quad h+1}{\bullet\!\!-\!\!-\!\!\bullet} \quad + \quad \sum_{\text{trees}} \quad \text{} \quad (8.25)$$

where now the sum runs over trees of the same type considered before but which no longer end up on the highest scale index. Furthermore, the last branches now represent $\mathcal{L}V^{(k)}$ if they are attached at scale k and the label $\mathcal{L}$ on the vertex nearest to the root in (8.25) indicates the action of $\mathcal{L}$. Also,

$$(1-\mathcal{L})V^{(h)} \quad = \quad \sum_{\text{trees}} \quad \text{} \quad (8.26)$$

So that we see that (8.25) provide us with a recursive calculation of $\mathcal{L}V^{(h)}$, giving $\vec{v}_h$, in terms of $\mathcal{L}V^{(h+1)}, \dots, \mathcal{L}V^{(N)} \equiv V^{(N)}$, i.e., in terms of a power series in $\vec{v}_{h+1}, \ldots, \vec{v}_N$. Also we see that (8.26) provides us with a power series expansion of $(1 - \mathcal{L})V^{(h)}$ in powers of $\vec{v}_{h+1}, \ldots, \vec{v}_N$.

What have we gained? Basically it turns out that the expansion (8.26) is much more regular than (8.20) in powers of $\vec{v}_N$ alone. The basic fact is that, if we assume that the running couplings are uniformly bounded by some ξ (for $d = 4$), then the sum of all the n-th order terms is bounded by $C_n = n!C^n\xi^n$ and the kernels $V(x_1, \ldots, x_p)$ can be bounded for all h as claimed in (6.16), provided one allows for a wider interpretation of the fields Φ appearing in (6.4).

To understand the latter observation, we can go back to the only explicit calculation of an irrelevant term done so far, namely (8.11). We see in that case that $(1 - \mathcal{L})\bar{V}$ naturally involves the fields (the $|y - x|^{-1}$ is added for convenience)

$$D^{\pm}_{yx\bar{\omega}} = \left(\psi^{\pm}_{y\bar{\omega}} - \psi^{\pm}_{x\bar{\omega}}\right) / |y - x| , \tag{8.27}$$

which it is convenient *not* to split and to allow it to be one of the Φ_z fields in (6.4), interpreting of course the z as a pair (x, y) of points and dz as $dxdy$.

A discussion of the details of the simple technical ideas involved in deriving the bounds can be found in §16 and 17 of [G1]; most of the key arguments are purely dimensional in nature and do not involve *hard* analytical estimates.

In all the above cases (infrared, ultraviolet, or Fermi liquid) one meets the same situation: the effective potential can be represented as in (6.4) or (6.5) by extending the meaning of Φ to cover a few (finitely many) more possibilities among which (8.27) is an example (see [G1], [BG1].

We now turn to (8.25), which gives us a recursive construction of the running constants $\vec{v}_h$. The ultraviolet case with $d = 3$ is somewhat different from the others because of the peculiarity already remarked; it is, however, easier to the point that it is one of the few cases in which the point of view that we are describing has been actually turned into proof (see [G1], [Rm1], [Rm2]). Therefore, we leave it aside to turn to the remaining more interesting cases ($\gamma > 0$, $d \geq 3$, $4 - d - 2\gamma \geq 0$ infrared, or $d = 4$ ultraviolet, or $d = 1$ Fermi liquid). The first basic result is a bound on the coefficients of the n-order term in the expansion (8.25). In all the above cases one finds that, provided $|\vec{v}_k| \leq \xi$ for all k's, the sum of the absolute values of the n-th order terms in the series (8.25) is bounded, for a suitable C, by

$$DC^{n-1}n!\,\xi^n \qquad \text{for all } k , \tag{8.28}$$

where D, C are suitable constants (depending only on the model and on the basic scale used in the multiscale analysis) but *not* on k. This is called the $n!$ bound, found in the remarkable work ([DCR]) on scalar φ_4^4 field theory (see also [G1], [BG1]).

The formal perturbation series in $v_{h+1}, \ldots, v_0$, defined in (8.25), is called the *beta functional* $B_h(v_{h+1}, \ldots)$.

Furthermore, if $4 - d - 2\gamma \geq 0, \gamma > 0$, infrared, or $d = 4$, ultraviolet, and in the Fermi liquid case, one can also show, <u>if $k > h$</u>, that $\vec{v}^{(k)}$ admit a power series expansion, in terms of $\vec{v}^{(h)}$ alone in the scalar cases and in terms of $\vec{v}^{(h)}$ and $\vec{v}^{(0)}$ in the case of the Fermi liquid, with coefficients again verifying bounds like (8.28) with an extra factor $(k-h)^n$ appearing (see [G1], §18–20; [BG1]). Note that the "extra factor" diverges as $k - h \to \infty$, making it impossible, unless the bound is improved, to make an expansion in terms of the bare constants.

This means that a formal power series links the running couplings $\vec{v}^{(h)}$ and $\vec{v}^{(h-1)}$ on neighboring scales; furthermore, the coefficients verify a uniform bound like (8.28) and, finally, they can be shown to have an order by order limit as $h \to -\infty$ or $N \to +\infty$ (depending on whether the problem is infrared or ultraviolet). One thus defines formal power series $B(\vec{v}), B'(\vec{v})$ such that

$$\vec{v}_{h-1} = B(\vec{v}_h) , \qquad \vec{v}_h = B'(\vec{v}_{h-1}) , \qquad (8.29)$$

which links, up to corrections vanishing as h becomes large, the running couplings on neighboring scales. The functions B and its inverse B', which are only formal power series, define the *beta function*.

If one could show that the series for B and B' converge or at least are asymptotically correct in describing the connection between $\vec{v}_h$ and $\vec{v}_{h-1}$, then we would have a method to control the $\vec{v}_h$ by studying the dynamical system

$$\vec{v}' = B(\vec{v}) . \qquad (8.30)$$

The models for which the B function exists as a well defined formal power series are called *renormalizable*.

The existence of the beta function, as a well-defined formal series, is the modern way to state the renormalizability of the model. We want to emphasize again that this procedure resums a huge number of Feynman graphs that are collected and summed to form the parameters v_h's (such values are very hard to compute and our real aim is not to really compute them). We will find that this can be possible if the *relation* (8.29) between these parameters is at least an *asymptotic series*.

Such information (or assumption) is sufficient in most cases to analyze perturbatively the *asymptotic freedom* and the possible *triviality* of the theory.

There are a few models for which the bound (8.28) can be improved by removing the $n!$. For such models, the beta functional is actually a holomorphic function of the parameters v_h near the origin (i.e., for $|v_h| < \xi$ for some $\xi > 0$ and all $h \leq 0$); in such cases one can really apply the program of the theory as we are developing it.

An example is the Fermi liquid beta functional which is in fact convergent at small $|v_h|$. An ultraviolet model for which the beta functional can be shown to exist and converge at small couplings is the $d = 2$ Gross-Neveu model [GK1]. The above convergence properties of the Fermi liquid beta functional rest upon the close connection between the Gross-Neveu model and the $d = 1$ Fermi liquid problem pointed out in [BG1].

Chapter 9
The Beta Function as a Dynamical System: Asymptotic Freedom of Marginal Theories

The use that one makes of the beta function is typically the following.

One forgets that the series defining it is only formal. One assumes that it is an asymptotic series to a well-defined function and, if so, one can argue that the sequence of running couplings behaves as prescribed by the approximations to B obtained by a truncation B^T of the beta function series. If all such approximations predict a bounded flow following suitable initial data, we have an algorithm to compute the $\vec{v}_h$ in the given approximation for B and, therefore, all the effective potentials as power series of the running couplings; such power series are free of divergences at any order.

This of course does not mean that the effective potentials are not singular in terms of the couplings $\vec{v}^{(0)}$ on scale $h = 0$ (usually called the *physical couplings*). It only says that we can hope that the effective potentials are analytic or C^{∞} as functions of all the running coupling $\vec{v}^{(h)}$; but it may well be that the structure of the dynamical system connecting the $\vec{v}^{(h)}$'s on neighboring scales is such that the couplings $\vec{v}^{(h)}$ as functions of $\vec{v}^{(0)}$ have singularities.

In other words, the above scheme, when the running couplings remain bounded uniformly, makes manifest that the effective potentials singularities as functions of the physical couplings are rather simple because they are due to singularities of the running couplings as functions of the physical ones, i.e., with respect to one among them (say $\vec{v}^{(0)}$).

To go beyond such results, one needs to understand precisely the convergence or analyticity properties of the beta function, which most of the time is an open and very difficult problem. But the progress achieved in understanding perturbation theory is in any case a major one.

We now apply the above program to the $d = 4$ ultraviolet problem and to the $4 - d - 2\gamma = 0$, $\gamma \geq 0$, $d > 3$ infrared problem as well as to the Fermi liquid using as B^T the simplest truncation, namely that to second order.

The calculation is elementary; one does not need the general tree formalism, which is necessary only to study the problem of the structure of the high orders.

The result is the following, for B:

$$\lambda' = \lambda - \beta_\gamma \lambda^2 + \beta'_\gamma \lambda \mu ,$$
$$\mu' = 2^{2-\gamma} \mu + \beta''_\gamma \lambda^2 + \beta'''_\gamma \lambda \mu + \beta''''_\gamma \mu^2 > 0, \ 4 - d - 2\gamma = 0 , \tag{9.1}$$

with $\beta_\gamma > 0$, and (see [G2], §20)

$$\lambda' = \lambda - \beta \lambda^2 + \beta' \lambda \mu ,$$
$$\mu' = 2\mu + \tilde{\beta} \lambda^2 + \tilde{\beta}' \mu^2 + \tilde{\beta}'' \lambda \mu + \tilde{\beta}''' \lambda \alpha , \tag{9.2}$$
$$\alpha' = \alpha + \beta'' \lambda^2 + \beta''' \mu^2 + \beta'''' \mu \alpha ,$$

for $d = 4$ ultraviolet and $d = 4$, $\gamma = 0$ infrared, with $\beta > 0$; while for the Fermi liquid ($d = 1$),

$$\lambda' = \lambda, \qquad \nu' = 2\nu, \qquad \zeta' = \zeta + \beta' \lambda^2, \qquad \alpha' = \alpha + \beta' \lambda^2 , \tag{9.3}$$

with $\beta' > 0$.

In the case (9.1) it is then easy to check, by using general methods of stability theory, that, given λ_0 is small enough, one can find $\mu_0 = O(\lambda_0^2)$ such that $\lambda_h \to 0, \mu_h \to 0$ as $h \to -\infty$ essentially as

$$\lambda_h \sim \frac{\lambda_0}{1 - \beta_\gamma h \lambda_0}, \qquad\qquad \mu_h \sim O(\lambda_h^2) . \tag{9.4}$$

The case (9.2), infrared, can also be treated. In this case we want to take $\alpha_0 = 0$ (as no gradient term is present at the beginning in the infrared problem). One easily finds that $\mu_0 = O(\lambda_0^2)$ can be found so that (9.2) generates a flow verifying (9.4) as well as

$$\alpha_h \longrightarrow \alpha_\infty = O(\lambda_0^2) \simeq \beta'' \sum_{h=0}^{-\infty} \left(\frac{\lambda_0}{1 - \beta h \lambda_0} \right)^2 \tag{9.5}$$

(and we see that α_∞ cannot be expected to be analytic at $\lambda_0 = 0$). So we conclude that, in the second-order approximation, the critical point exists and is a *mean field* critical point if $\gamma > 0$ (in the sense that the pair correlation behaves at large distance exactly as the free propagator). While the $d = 4$, $\gamma = 0$ critical point is a *mean field* model with, however, a *wave function* correction, i.e., at large distance the leading decay of the pair correlation is not equal to the free covariance but, rather, it is proportional to it with a constant of proportionality $(1 + \alpha_\infty)^{-1} < 1$.

The application of (9.2) to the $d = 4$ ultraviolet problem shows, of course, that it is not asymptotically free; one has to use (9.2) to evaluate λ from λ' (i.e., the inverse of the beta function) as we want to start from λ_0, μ_0, α_0 and construct λ_h, μ_h, α_h for large h. But no matter how small we start with λ_0, (9.2) will make λ_h grow to $O(1)$ at some finite h so that we are sure that we get out of a small neighborhood of the origin where the truncation approximation used for B is not justified. Similarly, one could try to start by taking a small $\vec{v}^{(N)}$. However, in this case the $v^{(h)}$, $h < N$, would have to be constructed by using (9.2) and this can be seen to imply that λ_0 will be $O(\lambda_N/(1 + \beta N \lambda_N)) \to 0$, $N \to \infty$.

The latter remark is at the origin of the well-known *triviality conjecture* of the quartic scalar model of QFT in $d = 4$. The status of this conjecture can be considerably improved ([Fr]), but the problem is wide open (see [GR]).

Coming to the Fermi liquid problem, we want to fix λ_0 small and show that ν_0 and α_0 can be fixed as functions of λ_0 so that taking $\zeta_0 = 0$ (as prescribed by the physical interpretation of the model), it is $\lambda_h \to 0$, $\nu_h \to 0$ (the latter to preserve the location of the Fermi surface), $\alpha_h - \zeta_h \to 0$, $\alpha_h, \zeta_h \to \zeta_\infty$ as $h \to \infty$. In this case we would have that the function $S(x)$ becomes at large distance proportional to $(1 + \zeta_\infty)^{-1} g(x)$ and we have a *normal Fermi* surface at p_F describing particles of mass m.

But it is clear that (9.3) is insufficient to draw such a conclusion. The second-order approximation does not allow us to decide whether λ_h grows or decreases. However, the fact that no λ^2 term is present means that, even if we go to higher order and eventually find that $\lambda_h \to 0$, the convergence to zero will not be very fast, so that $\sum_h \lambda_h^2 = +\infty$ (e.g., if the recursion was $\lambda' = \lambda - \beta\lambda^3$, then $\lambda_h \sim O(1/\sqrt{-h})$). In particular, the relations for ζ, α suggest that since $\sum \lambda_h^2 = \infty$ it would happen that α and ζ grow out of the domain where the truncation of B can be hoped to have a meaning.

This is a new situation very close to the one that one would meet in treating the infrared problem in $d = 3$, $\gamma = 0$ (which we have carefully avoided considering, so far (see chapter 12)). It is clear that in this case there seems to be a serious conceptual problem and perturbation theory does not seem possible in a consistent way. A new idea is necessary and in fact it was developed in the early days of renormalization group. We analyze it in the next chapters.

Chapter 10
Anomalous Dimension

The outcome of the discussion at the end of the last chapter, suggesting that $\alpha_h \to \infty$, can be interpreted as evidence that in the cases considered the scaling hypothesis is not correct and that at the critical point the asymptotic behavior of the pair Schwinger function is neither that of the potential nor that of the free propagator (corresponding to the trivial fixed point, $V = 0$, of $\mathcal{R}$).

We can extend the renormalization group methods so far followed to a more general approach which, when it works properly, would imply that the critical behavior of the pair Schwinger function is different from that of the free propagator.

We take as an example the $d = 3$, $\gamma = 0$ infrared problem, historically the first to be treated in this way ([WF]). But what we say applies word by word also to the theory of the Fermi surface, although we shall see that the results are quite different.

We start by setting $Z_0 = 1$ and by writing the effective potential integral (4.1) as

$$e^{-V_{\text{eff}}(\sqrt{Z_0}\varphi)} = \frac{1}{\mathcal{N}} \int P_{Z_0}^{(0)}(d\psi) e^{-V^{(0)}(\sqrt{Z_0}(\psi+\varphi))} , \qquad (10.1)$$

where $P_{Z_0}^{(h)}$ is the Gaussian measure with propagator with Fourier transform $Z_0^{-1}\Gamma_h(p)p^{-2}$ with

$$\Gamma_h(p) = e^{-(2^{-h}p/p_0)^2} . \qquad (10.2)$$

Then the field ψ can be represented as a sum of two independent fields,

$$\psi = \tilde{\psi}^{(0)} + \psi^{(<0)} , \qquad (10.3)$$

where the propagators for the two fields are, in order,

$$\frac{1}{Z_0}\tilde{g}^{(0)}(p) = \frac{1}{Z_0}\frac{\Gamma_0(p) - \Gamma_{-1}(p)}{p^2}, \qquad \frac{1}{Z_0}\frac{\Gamma_{-1}(p)}{p^2} , \qquad (10.4)$$

where the propagators are described symbolically by their Fourier transforms.

Therefore we can write

$$e^{-V_{\text{eff}}(\sqrt{Z_0}\varphi)} =$$

$$= \int P^{(0)}(d\tilde{\psi}^0)\, P^{(-1)}_{Z_0}(d\psi^{(<0)}) e^{-V^{(0)}(\sqrt{Z_0}(\tilde{\psi}^0+\psi^{(<0)}+\varphi))} = \qquad (10.5)$$

$$= \int P^{(-1)}_{Z_0}(d\psi^{(<0)}) e^{-\tilde{V}^{(-1)}(\sqrt{Z_0}(\psi^{(<0)}+\varphi))} \; ,$$

and $\tilde{V}^{(-1)}$ is defined by the same formulas describing $V^{(-1)}$ in the previous sections, as the $P^0(d\tilde{\psi}^0)$ integration is the same integral considered there.

Let $\mathcal{L}_3\tilde{V} = \tilde{\alpha}F_3$ be the projection of $\tilde{V}$ onto the direction $F_3(\psi) = \int dx : \psi_x(-\partial^2)\psi_x :$ obtained by applying the localization operator $\mathcal{L}$ in (8.2) and by selecting the part proportional to F_3. We can define

$$\mathcal{L}_3\tilde{V}^{(-1)}(\sqrt{Z_0}\,\psi) = \tilde{\alpha}_{-1}F_3(\sqrt{Z_0}\,\psi) \; ,$$

$$Z_{-1} = Z_0 + 2\tilde{\alpha}_{-1}Z_0 \; , \qquad (10.6)$$

$$\tilde{V}^{(-1)}(\sqrt{Z_0}\,\psi) = V^{(-1)}(\sqrt{Z_{-1}}\,\psi) + \tilde{\alpha}_{-1}F_3(\sqrt{Z_0}\,\psi) \; ,$$

where $V^{(-1)}$ is defined by the last relation, and it should not be confused with the $V^{(-1)}$ introduced in the previous sections (which coincides, at least in this first integration, with the above $\tilde{V}^{(-1)}$).

Using (10.6) and setting $D_0 = Z_{-1} - Z_0$, $\varphi \equiv \varphi_0$, we can write the exponential in (10.5), $e^{-V_{\text{eff}}(\sqrt{Z_0}\varphi)}$, as

$$c\int e^{-V^{(-1)}(\sqrt{Z_{-1}}(\psi^{(<0)}+\varphi_0))} e^{-\frac{1}{2}Z_0(\psi^{(<0)},p^2\Gamma^{-1}_{-1}\psi^{(<0)})} .$$

$$\cdot e^{-Z_0\tilde{\alpha}_{-1}(\psi^{(<0)}+\varphi_0,p^2[\psi^{(<0)}+\varphi_0])}d\psi^{(<0)} = \qquad (10.7)$$

$$= ce^{-\frac{1}{2}(D_0\varphi_0,p^2\varphi_0)}\int d\psi^{(<0)}\, e^{-\frac{1}{2}\left((Z_0\Gamma^{-1}_{-1}+D_0)\psi^{(<0)},p^2\,\psi^{(<0)}\right)} .$$

$$\cdot e^{-D_0(\psi^{(<0)},p^2\varphi_0)} e^{-V^{(-1)}\left(\sqrt{Z_{-1}}[\psi^{(<0)}+\varphi_0]\right)} =$$

$$= ce^{-\frac{1}{2}\left(D_0\varphi_0 - D_0^2(Z_0\Gamma^{-1}_{-1}+D_0)^{-1}\varphi_0,p^2\,\varphi_0\right)}\int d\psi\, e^{-\frac{1}{2}((Z_0\Gamma^{-1}_{-1}+D_0)\,p^2\psi,\psi)} .$$

$$\cdot e^{-V^{(-1)}\left(\sqrt{Z_{-1}}(\psi+\varphi_0 - D_0(Z_0\Gamma^{-1}_{-1}+D_0)^{-1}\varphi_0)\right)} \; ,$$

where the Gaussian integral over $\psi^{(<0)}$ has been formally written as an integral over $d\psi^{(<0)}$ times the exponential of the appropriate quadratic form, and c denotes the different (formal) normalizations.

Setting

$$\varphi_{-1} \equiv \frac{Z_0\Gamma_{-1}^{-1}}{Z_0\Gamma_{-1}^{-1} + D_0}\varphi_0 = \frac{1}{1 + D_0 Z_0^{-1}\Gamma_{-1}}\varphi_0 \,,$$

$$\frac{1}{Z_0\Gamma_{-1}^{-1} + D_0}\frac{1}{p^2} = \frac{1}{Z_{-1}}\tilde{g}^{(-1)}(p) + \frac{\Gamma_{-2}(p)}{Z_{-1}}\frac{1}{p^2}\,,$$

(10.8)

the Gaussian integral in the last step of (10.7) can also be thought of as an integral over two independent fields adding up to $\psi = \tilde{\psi}^{(-1)} + \psi^{(<-1)}$ with propagators given by the last two terms in the second of equations (10.8). Hence,

$$e^{-V_{\text{eff}}(\sqrt{Z_0}\,\varphi_0)} =$$
$$= c\int e^{-V^{(0)}\left(\sqrt{Z_0}\,(\tilde{\psi}^{(0)}+\psi^{(<0)}+\varphi_0)\right)} P^{(0)}(d\tilde{\psi}^{(0)})\,P_{Z_0}^{(-1)}(d\psi^{(<0)}) =$$
$$= c\,e^{-\frac{1}{2}D_0(\varphi_{-1},p^2\varphi_0)}.$$

$$\cdot \int e^{-V^{(-1)}\left(\sqrt{Z_{-1}}\,(\tilde{\psi}^{(-1)}+\psi^{(<-1)}+\varphi_{-1})\right)} P^{(-1)}(d\tilde{\psi}^{(-1)})\,P_{Z_{-1}}^{(-2)}(d\psi^{(<-1)})\,,$$

(10.9)

which can be easily iterated to yield

$$e^{-V_{\text{eff}}(\varphi_0)} = c\,e^{-\frac{1}{2}\sum_{j=h}^{-1}D_{j+1}\left(\varphi_j,p^2\varphi_{j+1}\right)}.$$

$$\cdot \int e^{-V^{(h)}\left(\sqrt{Z_h}\,(\tilde{\psi}^{(h)}+\psi^{(<h)}+\varphi_h)\right)} P^{(h)}(d\tilde{\psi}^{(h)})P_{Z_h}^{(h-1)}(d\psi^{(<h)})\,,$$

(10.10)

with $D_j = Z_{j-1} - Z_j$, and

$$\varphi_h = \prod_{j=h+1}^{0}\frac{1}{1 + \left(\frac{Z_{j-1}}{Z_j} - 1\right)\Gamma_{j-1}(p)}\varphi_0 \,,$$

$$\frac{1}{Z_h}\tilde{g}^{(h)}(p) \equiv \left(\frac{1}{Z_{h+1}\Gamma_h^{-1}(p) + D_{h+1}} - \frac{1}{Z_h\Gamma_{h-1}^{-1}(p)}\right)\frac{1}{p^2}\,,$$

(10.11)

and

$$V^{(h)}(\sqrt{Z_h}\,\psi) =$$
$$= -(1 - \mathcal{L}_3)\log\int P^{(h+1)}(d\tilde{\psi}^{(h+1)})\,e^{-V^{(h+1)}(\sqrt{Z_{h+1}}\,(\tilde{\psi}^{(h+1)}+\psi))} \equiv$$
$$\equiv (1 - \mathcal{L}_3)\tilde{V}^{(h)}(\sqrt{Z_{h+1}}\,\psi)\,,$$

(10.12)

$$Z_h = Z_{h+1} + 2\tilde{\alpha}_h Z_{h+1} \quad \text{if} \quad \mathcal{L}_3\tilde{V}^{(h)}(\psi) = \tilde{\alpha}_h F_3(\psi)\,.$$

One can check that, if $h \leq -1$,

$$p^2 \tilde{g}^{(h)}(p) = \left(\Gamma_h(p) - \Gamma_{h-1}(p)\right) + \Gamma_h(p)\left(1 - \Gamma_h(p)\right)\frac{z_h}{1 + z_h\Gamma_h(p)} , \quad (10.13)$$

if $z_h = (Z_h - Z_{h+1})Z_{h+1}^{-1}$. Therefore, if z_h is asymptotically constant, say if $z_h = 2^{2\eta} - 1 + O(2^h)$ for some η (i.e., if $Z_h \sim 2^{-2\eta h}$), the fields $\tilde{\psi}^{(h)}$ have essentially the same distribution as the fields

$$\frac{2^{\delta h}}{\sqrt{Z_h}}\overline{\psi}_{2^h.} , \quad (10.14)$$

where $\delta = \frac{d-2}{2}$ and $\overline{\psi}$ has propagator $\tilde{g}$ calculated from (10.13), with $h = 0$ and z_0 replaced by $2^{2\eta} - 1$.

It is therefore convenient to introduce the *dimensionless anomalous effective potential* on scale h (see (6.1)) by

$$\overline{V}^{(h)}(\psi.) = V^{(h)}(2^{\delta h}\psi_{2^h.}) \quad (10.15)$$

so that the recursion relation becomes

$$\overline{V}^{(h-1)}(\psi) =$$
$$= -(1 - \mathcal{L}_3) \log \int P(d\overline{\psi}) \exp -\overline{V}^{(h)}(\overline{\psi} + \sqrt{\frac{Z_h}{Z_{h-1}}} 2^{-\delta}\psi_{2^{-1}.}) \quad (10.16)$$

and the integral over $\overline{\psi}$ is a Gaussian integral with propagator (10.13) with $\Gamma_h(p)$ replaced by $\Gamma_0(p)$, and Γ_{h+1} by Γ_1. The equation (10.16) has to be complemented by the definition of Z_h; namely, $Z_{h-1} = Z_h + 2\tilde{\alpha}_h Z_h$ with $\tilde{\alpha}_h Z_h/Z_{h-1}$ being the coefficient of F_3 in $\mathcal{L}_3\tilde{V}$ if $\tilde{V}$ is the argument of the logarithm in (10.16).

If $\frac{Z_h}{Z_{h-1}}$ can be regarded as essentially proportional to $2^{-2\eta}$, the relation (10.16) becomes (essentially) the scale-independent map,

$$\overline{V}'(\psi) = -(1 - \mathcal{L}_3) \log \int P(d\overline{\psi}) \exp -\overline{V}(\overline{\psi} + 2^{-\delta^*}\psi_{2^{-1}.})$$
$$\equiv \mathcal{R}_\eta \overline{V} , \quad (10.17)$$

where $\delta^* \equiv \delta + \eta$ and the $\overline{\psi}$ Gaussian integration is with propagator $\overline{g}(p)$:

$$\overline{g}(p) = \frac{1}{p^2}\left(\Gamma_0(p) - \Gamma_{-1}(p)\right) +$$
$$+ \Gamma_0(p)\left(1 - \Gamma_0(p)\right)\frac{2^{2\eta} - 1}{1 + (2^{2\eta} - 1)\Gamma_0(p)} . \quad (10.18)$$

Of course, the value of η is not a prioriknown and therefore one should still use the map (10.16) with the accompanying definition of Z_h in order to obtain the sequence of $V^{(h)}$.

By the methods of the previous sections one establishes the existence of a beta functional with the running couplings $v_h = (\lambda_h^{(6)}, \lambda_h^{(4)}, \mu_h)$ being related to the coefficients $\lambda_h^{(6)}, 2^h \lambda_h^{(4)}, 2^{2h} \mu_h$ of $F_6(\sqrt{Z_h}\,\psi)$, $F_4(\sqrt{Z_h}\,\psi)$ and $F_2(\sqrt{Z_h}\,\psi)$, i.e., of the relevant and marginal operators: note that by the construction *no operator F_3 will ever appear*. Furthermore, the beta functional will depend on the ratios Z_q/Z_{q-1}, $q \geq h$, which in turn are determined recursively (i.e., in some sense the sequence of missing running couplings has been replaced by the sequence Z_h/Z_{h-1}).

One finds a relation such as

$$v_{h-1} = M_h v_h + B_h\left(v_h, v_{h+1}, \ldots, v_0; \frac{Z_{h+1}}{Z_h}, \ldots\right),$$

$$1 = \frac{Z_h}{Z_{h-1}}\left(1 + A_h\left(v_h, v_{h+1}, \ldots, v_0; \frac{Z_{h+1}}{Z_h}, \ldots\right)\right), \tag{10.19}$$

where the A_h, B_h are formal power series in their arguments v_j, and M_h is a diagonal matrix with diagonal elements ≥ 1 if $Z_h/Z_{h-1} \geq 1$. More precisely the diagonal elements are, respectively,

$$\left(\frac{Z_h}{Z_{h-1}}\right)^3 2^{6-2d}, \quad \left(\frac{Z_h}{Z_{h-1}}\right)^2 2^{4-d}, \quad \left(\frac{Z_h}{Z_{h-1}}\right)^2 2^2, \tag{10.20}$$

with $d = 3$, in the present case.

One can prove that if

$$\left|\sqrt{\frac{Z_h}{Z_{h-1}}}\right| < \xi \tag{10.21}$$

for some $\xi > 0$, and for all $h \leq 0$, then the coefficients of the expansions in the v's verify bounds such as

$$|C_{\vec{n}}| \leq 2^{a|\vec{n}|} 2^{-bs(\vec{n})} |\vec{n}|! D \tag{10.22}$$

if $\vec{n} = \{n_1, \ldots, n_{|h|}\}$ and $C_{\vec{n}} = $ (coeff. of the monomial $v_{h+1}^{n_1} v_{h+2}^{n_2} \cdots v_0^{n_{|h|}}$), and $s(\vec{n}) = k - h$ if $n_{k+1} = n_{k+2} = \ldots = n_{|h|} = 0$ but $n_k > 0$; the a, b, D are positive constants dependent on ξ but *independent of h*.

The term 2^{-bs} shows that the beta functional has *short memory* as it means that A_h, B_h depend only very slightly on the running couplings of order k large compared to h.

Likewise, the dimensionless effective potentials are expressed as

$$\overline{V}^{(h)}(\varphi) = \sum \int \overline{V}^{(h)}(x_1, \ldots, x_m) \prod_{i=1}^{m} \partial^{\mu_j} \varphi_{x_i}^{\nu_i} , \qquad (10.23)$$

with $\nu_i \leq 6$ (6 being the maximal degree of the relevant and marginal operators) nonnegative integers and $\mu_j \leq 2$; and the kernels $V(\cdot)$ admit an expansion in the running constants $v_{h+1}, \ldots, v_0$ with coefficients depending on the ratios Z_q/Z_{q-1} with $q > h$; and if the latter are bounded as in (10.21), they verify the bounds similar to (10.22); with the notations of (10.22) the bounds are, for $\kappa > 0$ suitable,

$$|\overline{V}_{\vec{n}}^{(h)}(x_1, \ldots, x_m)| \leq 2^{a|\vec{n}|} 2^{-bs(\vec{n})} |\vec{n}|!^2 \, D \, e^{-\kappa d(x_1 \ldots, x_m)} , \qquad (10.24)$$

with $d(x_1, \ldots, x_m)$ being the length of the shortest graph connecting the points $x_1, \ldots, x_m$.

The (10.23), (10.24) are a convenient way of expressing the $\overline{V}^{(h)}$ and its qualitative properties. However, the proof of (10.24) leads to an expression of $\overline{V}^{(h)}$ of somewhat different and more involved form, implying the (10.23), quite naturally associated with the theory's Feynman graphs. Such a "more natural" expression is briefly summarized in appendix 4.

The bound (10.24) and the equation (10.23) make it manifest that the functional derivative of $\overline{V}^{(h)}$ have kernels verifying the same bounds, e.g., if we write

$$\frac{\delta \overline{V}^{(h)}(\varphi)}{\delta \varphi_x} = \sum \int \overline{V}_x^{(h)}(x_1, \ldots, x_m) \prod_{i=1}^{m} \partial^{\mu_j} \varphi_{x_i}^{\nu_i} , \qquad (10.25)$$

the kernels $\overline{V}_x^{(h)}$ verify

$$|\overline{V}_{x;\vec{n}}^{(h)}| \leq 2^{a|\vec{n}|} 2^{-bs(\vec{n})} |\vec{n}|!^2 \, D \, e^{-\kappa d(x, x_1 \ldots x_m)} . \qquad (10.26)$$

Note that this bound is not a trivial consequence of (10.24), since the functional derivative of $\partial \varphi_y$ with respect to φ_x is the derivative of the delta function $\delta(x - y)$, and, therefore, a bound on the functional derivatives of $\overline{V}^{(h)}$ involves a bound on the smoothness properties of the kernels in (10.23) (see [BGPS]).

This shows that if one can find an initial $V^{(0)}$ so that (10.21) holds and $|v_h| < \varepsilon$ for some $\xi, \varepsilon, \kappa > 0$ independent on $h \leq 0$, then a perturbation theory in terms of the running couplings for $\overline{V}^{(h)}$ can be formally constructed.

Alternatively, one can look for a *nontrivial* fixed point of the anomalous renormalization transformation, i.e., for a pair v, η and for a suitable matrix M_η, such that

$$v = M_\eta v + \lim_{h \to -\infty} B_h(v, v, .., v; 2^{-2\eta}, 2^{-2\eta}, ..) = M_\eta v + B(v, \eta) \ ,$$

$$1 = 2^{-2\eta}\big(1 + \lim_{h \to -\infty} A_h(v, v, .., v; 2^{-2\eta}, 2^{-2\eta}, ..)\big) = \qquad (10.27)$$

$$= 2^{-2\eta}(1 + A(v, \eta)) \ ,$$

which, at least if A, B are supposed to admit an asymptotic expansion in powers of v, η, can be studied by series expansion. We discuss this point in some detail in chapter 12.

If a solution v^*, η^* to (10.27) can be found, then one can consider the effective potential V^* defined by setting all v_j equal to v^* and all the ratios Z_h/Z_{h-1} equal to $2^{-2\eta^*}$ in the series defining the kernels for the $V^{(h)}$'s, for $h \to -\infty$. It is very remarkable that V^* will, by construction, be such that

$$\mathcal{R}_{\eta^*} V^* \equiv V^* \ , \qquad (10.28)$$

i.e., it is a (formal) fixed point for the anomalous renormalization transformation.

If the V^* is an unstable fixed point, with a one-dimensional instability (i.e., if the linearization of the anomalous $\mathcal{R}_{\eta^*}$ transformation around V^* has only one eigenvector with eigenvalue > 1), then by changing the inverse temperature β we can hope that the curve described by $V^{(0)}$ crosses the stable manifold of V^* for some value β_c. At such temperature, the asymptotic behavior of the pair Schwinger function will be described by V^* rather than by the trivial fixed point $V = 0$ (as in the cases with $d > 4$, or even $d = 4$ and $\gamma = 0$, where the trivial fixed point has a one-dimensional instability).

It is not difficult to give heuristic arguments showing that the asymptotic behavior of the pair Schwinger function will be *anomalous*, in the circumstances considered in the last comment, in the sense that at $\beta = \beta_c$ the behavior will be $\sim |x - y|^{-(d-2+2\eta^*)}$ instead of the *trivial* $\sim |x - y|^{-(d-2)}$ (see [BG1], [BGM]).

To obtain a more precise statement it is necessary to find an expression of the Schwinger functions in terms of the anomalous effective potentials. That the problem is somewhat tricky can be immediately realized from (10.1), (10.10) and from the general relation (4.3), which would in fact lead to expressing the generating functional for the Schwinger functions $S^T_{int}(\varphi)$ as something like

$$S^T_{int}(\varphi) = \lim_{h \to -\infty} \qquad (10.29)$$

$$\frac{1}{2}\left[\left(\frac{\Gamma_0}{Z_0 p^2}\varphi, \varphi\right) - \sum_{j=h}^{-1} D_{j+1}\left(\varphi_j, \frac{\Gamma_0^2}{Z_0^2 p^2}\varphi_{j+1}\right)\right] - V^{(h)}\left(\sqrt{Z_h}\, \frac{\Gamma_0}{Z_0 p^2}\varphi_h\right),$$

where $\varphi = \varphi_0$ and φ_j, $j < 0$, is defined by (10.11).

But the equation (10.29) does not immediately make sense at $h = -\infty$ because, for instance, the series is manifestly divergent. However, a simple heuristic argument suggests that to study the S^T_{int} the h in (10.29) should not be sent to $-\infty$ but rather it should be fixed so that $2^h p_0$ is the smallest momentum scale in the Fourier transform of φ (i.e., a value below which the Fourier transform of φ is essentially zero), which can be interpreted as an infrared cutoff. Hence, one expects that, if $p \sim 2^{\bar h} p_0$, the Fourier transform $\hat S_2(p)$ of the pair Schwinger function behaves as

$$\frac{1}{p^2}\left[\frac{1}{Z_0} - \sum_{j=h}^{-1} \frac{D_{j+1}}{Z_j Z_{j+1}}\right] = \frac{1}{Z_h p^2}. \qquad (10.30)$$

In fact, if $h \gg \bar h$, the first of (10.11) implies that $\varphi_j \simeq \varphi/Z_j$; it follows that, if $Z_h \sim 2^{-2\eta h}$, $\hat S_2(p) \sim p^{-(2-2\eta)}$, implying that $S_2(x - y) \sim |x - y|^{-(d-2+2\eta)}$ for $|x - y| \to \infty$.

It is, however, not easy to make the latter statement more precise and it is more convenient to find an expression for the Schwinger functions generating functional S^T_{int}, which depends on the $V^{(h)}$ and which involves no divergent pieces. This can be achieved under the sole assumption that the dimensionless potentials verify the bounds (10.24), (10.26) (see [BGPS]).

The conclusion is that under the hypothesis that the latter bounds hold and that $Z_h \sim 2^{-2\eta h}$, the pair Schwinger function $S(x - y)$ can be shown to have the asymptotic behavior

$$S(x - y) = \sum_{h=-\infty}^{0} \frac{1}{Z_h} \tilde g^{(h)}(x - y)\left(1 + O(\max_h |v_h|^2)\right), \qquad (10.31)$$

provided $\mu_h \xrightarrow[h \to -\infty]{} 0$. The (10.31) is easily seen to imply that S behaves at ∞ as $|x - y|^{-(d-2+2\eta)}$ (see [BGPS] for a detailed analysis). The $\mu_h \to 0$ as $h \to -\infty$ is the natural condition that can be imposed to determine the initial $V^{(0)}$, i.e., the critical temperature.

Therefore, it is natural to try to see if there is another *nontrivial* fixed point of the map $\mathcal{R}_\eta$. We do not examine here the evidence for the existence of such a fixed point. We point out that just the possibility of performing the above analysis and of discovering the possible mechanism for the development of an anomalous dimension in the long-range behavior at the critical point has been an important success of the renormalization group methods of analysis [WF] and, even if one regards the analysis as purely heuristic, it has to be stressed that *the possibility of nonclassical critical indices (i.e., of nonzero anomaly η) is probably the most important achievement of the renormalization group.*

The analysis can be repeated in the case of Fermi liquids. One has, however, to decide which of the two gradient terms one should eliminate at every step from the effective potential: for instance, one can decide to eliminate the appropriate part of ζF_4 (or equivalently that of αF_3), because now $\alpha \neq \zeta$ and one has here an arbitrary choice. Eliminating ζF_4 and calling now δ_h the coefficient of F_3, the final result for the pair Schwinger function is similar to (10.31), and it has the same interpretation

$$S(x - y, \vec{\omega}) = \frac{1}{(x_0 + i\vec{\omega} \cdot \vec{x})(x_0^2 + \vec{x}^2)^\eta} \cdot (1 + O(\max_h |v_h|^2)) \,, \quad (10.32)$$

if $v_h, \delta_h \to 0$; so we see that $v_h, \delta_h \to 0$ are the natural conditions to determine v_0, α_0 in terms of λ_0.

In the physics literature, $\Gamma_h(p)$ is often replaced by the characteristic function $\chi(|2^h p| < p_0)$ and the formal analysis becomes clearer in some respects (and obscure in others). An example of such analysis can be found in chapter 15.

We conclude this chapter by a rather simple remark: it is clear that the formal power series in $v_h, \ldots, v_0$ can be rewritten, by formal series manipulations, in the form

$$\begin{aligned}
v_{h-1} &= M_h v_h + \hat{B}_h(v_h, \frac{Z_{h+1}}{Z_h}) \,, \\
1 &= \frac{Z_h}{Z_{h-1}} (1 + \hat{A}(v_h, \frac{Z_{h+1}}{Z_h})) \,,
\end{aligned} \quad (10.33)$$

and it is interesting to note that the bounds (10.22) together with the fact that the diagonal matrix elements of M_h are ≥ 1, if $Z_h/Z_{h-1} > 1$, imply that the coefficients of $\hat{B}, \hat{A}$ in the v_h can be bounded by

$$|a_{h,n}|, |b_{h,n}| < \hat{D}\hat{C}^n n! \,, \tag{10.34}$$

with $\hat{D}, \hat{C}$ computable in terms of a, b, D in (10.22).

This is done by first eliminating the ratios Z_p/Z_{p-1} obtaining a recursion relation for the v_h having the form $v_{h-1} = M_h v_h + \tilde{B}_h(v_h, \ldots, v_0)$, with the $\tilde{B}_h$ given by formal power series verifying the same bounds as the B_h and, subsequently, by expressing v_{h+1} in terms of v_h, and $v_{h+2}, \ldots, v_0$ by inverting

$$v_{h+1} = M_{h+1}^{-1} v_h - M_{h+1}^{-1} \tilde{B}_h(v_{h+1}, \ldots, v_0) \,, \tag{10.35}$$

and then v_{h+2} in terms of $v_h, v_{h+3}, v_{h+4}, \ldots$ and so on until all the v_j, $j > h$ are eliminated.

One can show that the formal functions $\hat{B}_h(v, w), \hat{A}_h(v, w)$ thus obtained have a limit as $h \to -\infty$ order by order in v and for each w. The formal limits $\hat{A}(v, w), \hat{B}(v, w)$ define the *scaling beta function*, and the formal functions $\hat{A}_h, \hat{B}_h$ define the *beta function*.

It should be noted, however, that the beta functional is a better-defined object, as one can see that in some simple cases the series in the many variables v_h defining it are convergent if, for a small enough $\xi > 0$, $|v_h| < \xi$ for all $h \leq 0$. But in the same cases the beta functions defining formal series are very probably not convergent.

This raises a very interesting problem: Does there exist a family of functions $\hat{B}_h, \hat{A}_h$ such that the running couplings can be defined by the dynamical system described by (10.33) starting from suitably restricted initial v_0? And admitting an asymptotic series at the origin given by the formal power series defined above?

In most works on the renormalization group such an assumption is tacitly or explicitly made, although there seems to be no general *non-perturbative* definition of the beta functions (in the models that we are considering). This is not very visible, usually, because the use that is made of the beta function is to regard as a good approximation to it the second-order truncation of the formal series defining them, which, by the way, coincides with the second order truncation of the corresponding beta functionals. Nevertheless, this constitutes one of the outstanding problems of the theory — a problem for which there seems to be not only no solution in sight, but not even ideas toward obtaining a solution.

Chapter 11
The Fermi Liquid and the Luttinger Model

We have seen that the beta function for this model, discussed in chapter 9, seems inconsistent with the normal scaling, i.e., with an asymptotic behavior of the interacting propagator equal or proportional to that of the free propagator.

Note that this is true in spite of the fact that the beta functional is given by a convergent series for small couplings. Thus this shows that it is possible that the beta functional exists not only formally but as a convergent power series in the running couplings, but, nevertheless, it does not help in solving the problem because it generates a flow leading away from the neighborhood of the origin where the functional makes sense.

We attribute the above accident to the fact that the scaling is anomalous, i.e., that the pair Schwinger function behaves at ∞ differently from the free one. Hence we can try to apply the scheme devised in the previous section to discuss the possibility of an anomalous scaling.

The effective potentials now have a different definition, and basically at every application of R one extracts a part of the newly created effective potential and puts it in the Grassmannian integration, as described in deriving the recursion relation (10.10).

We choose to extract the part with the $\int : \psi^+_{x\vec{\omega}} \partial_t \psi^-_{x\vec{\omega}'} : e^{i(\vec{\omega}-\vec{\omega}')\cdot\vec{x}p_F}$ operator (we could also choose $\int : \psi^+_{x\vec{\omega}_1}\vec{\omega}_2 \cdot D_{\vec{\omega}_2} \psi^-_{x\vec{\omega}_2} : e^{i(\vec{\omega}_1-\vec{\omega}_2)\vec{x}p_F}$), i.e., we try to define the renormalization transformation so that $\zeta_h \equiv 0$ (in the example of chapter 10 we imposed $\alpha_h = 0$). From the point of view of the theory of the beta functional, one has to do essentially the same analysis and the above is a minor modification.

One finds that, calling δ_h the coefficient of F_3 and $v_h = (\lambda_h, \delta_h, \mu_h)$, the beta functional can be written as the recurrence relation,

$$v_{h-1} = \Lambda v_h + B(v_h) + \mathcal{B}_h(v_h, v_h - v_{h+1}, v_h - v_{h+2}, \ldots, v_h - v_0) , \quad (11.1)$$

with Λ linear and diagonal, and $B, \mathcal{B}_j$ analytic in their arguments in a polydisk of radius ε_0. The latter is a remarkable property that is a consequence of the fermionic nature of the model and can be proved by using the same ideas that lead to the theory of the two-dimensional

Gross–Neveu model ([GK1]). The ratios Z_h/Z_j, $j > h$ that one would expect to appear in the r.h.s. have been eliminated by using the definition of Z_{h-1}/Z_h that comes from the relation analogous to the second of (10.19).

Furthermore, $\mathcal{B}_j$ are functions with *short memory*; this means that functions $D_j^k(\vec{x}_k, \vec{x}_{k+1}, \ldots, \vec{x}_{-1}, \vec{x}_0)$ exist such that for some $b, d > 0$,

$$\mathcal{B}_j(\vec{x}_j, \vec{x}_{j+1}, \ldots, \vec{x}_{-1}, \vec{x}_0) = \sum_{k=j}^{-1} D_j^k(\vec{x}_j, \vec{x}_{k+1}, \ldots, \vec{x}_{-1}, \vec{x}_0) \,,$$

$$|D_j^k| \leq 2^{-(k-j)b} |\vec{x}_j| \left(\sup_{k'>k} |\vec{x}_{k'}| \right) d, \qquad k \geq j \,. \tag{11.2}$$

The exponential decay of the D-functions has the practical consequence that the dynamical system without memory defined by the map, $v' = \Lambda v + B(v)$, and the one in (11.1) behave essentially in the same way near $v_h = 0$.

The *important* part of (11.1) is the one that does not tend to zero because $h \to -\infty$, when the x_h have a limit; i.e., it is the one obtained by setting $\mathcal{B}_j \equiv 0$.

The dynamical system $v' = \Lambda v + B(v)$ has properties depending critically on the function $G(\lambda, \delta) \equiv B(\lambda, \delta, 0)$.

For instance, if we look for a nontrivial fixed point $v = v + B(v)$ with $v = (\lambda, 0, 0)$, we see that one exists if $G(\lambda, \delta) = 0$ has a solution.

Assuming $\lambda^3 G(\lambda, 0) = \beta \lambda^p + \beta' \lambda^{p+1} + \ldots$, for some $p \geq 0$ we find, by a slightly more detailed analysis of the structure of the functions B, that there are trajectories in which $\delta_h, \nu_h \to 0$ as $h \to -\infty$.

On such trajectories $\lambda_h \to 0$ for $h \to -\infty$, if $\beta \lambda_0^{p-1} < 0$ but very slowly (like $1/|h|^{1/(p-1)}$). The last of (10.19), neglecting terms that tend to zero as $O(2^h)$ for $h \to -\infty$, in the present case becomes

$$\begin{aligned} 1 = &\frac{Z_h}{Z_{h-1}} \Big[1 + \lambda_h^2 B^1(\lambda_h, \lambda_{h+1}, \ldots, \lambda_0) + \\ &+ \delta_h \lambda_h^2 B^2(\lambda_h, \delta_h, \nu_h, \ldots, \lambda_0, \delta_0, \nu_0) + \\ &+ \lambda_h^2 \nu_h^2 B^3(\lambda_h, \delta_h, \nu_h, \ldots, \lambda_0, \delta_0, \nu_0) \Big] \,, \end{aligned} \tag{11.3}$$

with $B^1(0, 0, \ldots, 0) \neq 0$. And in the latter case, still neglecting for consistency the terms tending to zero for $h \to -\infty$, yields

$$\log Z_h \propto \sum_{h'>h} |h'|^{-2(p-1)^{-1}} \simeq |h|^{1-2(p-1)^{-1}} \,, \tag{11.4}$$

hence Z_h *does not* have the form $2^{-h\eta}$ for some $\eta > 0$.

This is the remark that provides us a key to the proof that $G = 0$. To study the latter conjecture we can try to look for a model, possibly *totally* different from our Fermi liquid, for which

1. one can describe it by a renormalization group method;
2. the beta functional of the model has the same form as (11.2) with possibly different D functions;
3. $G(\lambda)$ is the same for both models;
4. one knows that the model has $\eta(\lambda_0) = \bar{\beta}\lambda_0^2 + O(\lambda^3)$.

In [BG1] a simple proof is presented showing that the Luttinger model meets the requirements of (1), (2), (3). It is also known from the exact solution of the Luttinger model ([L]) by Mattis and Lieb ([ML], [LM]) that the propagator decays at ∞ faster than the free propagator by a factor $\propto |x - y|^{-\eta(\lambda_0)}$, $\eta(\lambda_0) = \lambda_0^2 \bar{\beta} + \dots$.

In [BGPS] it is shown that the converse, $G(\lambda, \delta) \neq 0$, is incompatible with the known asymptotic behavior of the exact solution of the Luttinger model. Therefore it follows that $G(\lambda, \delta)$ must be zero.

The $\eta \neq 0$ intuitively corresponds to a singularity at the Fermi surface of the Fourier transform of the propagator that is weaker than the usual jump but is described by a vertical slope such as $||\vec{k}| - p_F|^{2\eta}$, $\eta = \bar{\beta}\lambda_0^2 + \dots$.

If $G(\lambda, \delta) = 0$, the analysis of the flow (11.1) is easy, and we see that one can always fix δ_0 and ν_0 (i.e., α_0 and ν_0) so that $\alpha_h, \nu_h \to 0$ and $\lambda_h \to \lambda_{-\infty}$ ([BGPS]).

In this case the heuristic arguments leading to (10.32) can be made rigorous and the (11.3) provides us with an asymptotic limit for $2^{2\eta} = \lim_{h \to -\infty} Z_{h-1}/Z_h$ and $\eta = O(\lambda_{-\infty}^2)$, thus determining an "anomalous" asymptotic behavior of the pair Schwinger function.

More generally, given λ_0 and α_0, one can fix $\nu_0 = \nu_0(\lambda_0, \delta_0)$ so that the pair Schwinger function behaves asymptotically as in (10.32), with a coefficient $\beta(\lambda_0, \alpha_0) \neq 1$ in front of $i\vec{\omega} \cdot \vec{x}$.

The above properties can also be qualitatively stated as follows: the spinless, $d = 1$, Fermi liquid is *well* described by the Luttinger model, as far as the interacting propagator is concerned, and it corroborates the pioneering work of Tomonaga ([T]).

We finally remark that the fact that $v_h \to v_{-\infty}$ and $Z_{h-1}/Z_h \to 2^{2\eta}$ can be interpreted as saying that the recursion relation for the dimensionless anomalous effective potential introduced in (10.15) and corresponding to the sequence of the running couplings $v_h \equiv v_{-\infty}$,[1], evaluated at scale $h \to -\infty$ is a fixed point for the anomalous recursion relation (10.17).

Hence the asymptotic behavior of the renormalization group flow is controlled by a two-dimensional continuum of nontrivial fixed points parameterized by δ_0, λ_0 [2], and determined by the exact solutions of the Luttinger model (because the leading part of the beta function for the Luttinger model and that for the class of models considered coincide).

---

[1] Recall that once the sequence of the running couplings v_h is given and they are sufficiently small, one can reconstruct the full effective potential by a convergent series expansion in the sequence.

[2] But the δ_0 plays a somewhat trivial role and disappears if one decides to fix the Fermi velocity at 1, hence sometimes one says that there is a "line of fixed points."

Chapter 12
The Generic Critical Point for $d=3$, $\gamma=0$: The ε-Expansion

The Fermi liquid model (with $d = 1$) is a rather special example of anomalous scaling (as the beta function vanishes). A more classical but somewhat more complicated example is provided by the statistical mechanics of three-dimensional spin systems: this is the $d = 3, \gamma = 0$ case of chapter 2, for which we have not yet reached any conclusions.

In this case the relevant operators are all the four operators in (6.14), and we thus have four running couplings, which we denote μ_h (for $: \varphi^2 :$), α_h (for $: (\partial\varphi)^2 :$), $\lambda_h^{(4)}$ (for $: \varphi^4 :$) and $\lambda_h^{(6)}$ (for $: \varphi^6 :$): $\vec{v}_h = (\lambda_h^{(6)}, \lambda_h^{(4)}, \mu_h, \alpha_h)$.

We can easily compute the beta function to second order as (writing only a selected few of the second order terms)

$$
\begin{aligned}
\lambda_{h-1}^{(6)} &= \lambda_h^{(6)} - \beta\lambda_h^{(6)2} - \beta'\lambda_h^{(4)2} + \dots , \\
\lambda_{h-1}^{(4)} &= 2\lambda_h^{(4)} + \dots , \\
\mu_{h-1} &= 4\mu_h + \dots , \\
\alpha_{h-1} &= \alpha_h + \beta''\lambda^{(6)2} + \dots ,
\end{aligned}
\tag{12.1}
$$

where $\beta, \beta', \beta'' > 0$ and no α_h^2 terms appear in the last recursion, as well as no terms that are linear in α_h besides the one involving $\alpha_h\mu_h$.

We are interested, when studying the φ^4 model of chapter 2, to start with (see (7.2))

$$
\begin{aligned}
\lambda_0^{(6)} &= 0 , & \lambda_0^{(4)} &= L_0\beta^{-2} , \\
\alpha_0 &= 0 , & \mu_0 &= 6L_0\beta^{-2}C_{xx} + (-R_0\beta^{-1} + r) .
\end{aligned}
\tag{12.2}
$$

We see that, already in the linear approximation, we cannot keep $\lambda_h^{(4)}$ bounded because it is now amplified by a factor 2 in each iteration.

If we had $\lambda_0^{(6)} > 0$ and $\lambda_0^{(4)}$ as a free parameter we could think of fixing $\lambda_0^{(4)}, \mu_0$ as functions of $\lambda_0^{(6)}$ so that the running couplings had a bounded flow. To first order we would have to take $\lambda_0^{(4)} = \mu = 0$. To second order (or higher) we would have to take $\lambda_0^{(4)}, \mu_0$ as functions of $\lambda_0^{(6)}$ so that the running couplings stay bounded.

Thus we see that a general φ^6-*model*, i.e., a model with high temperature distribution (see chapter 2) defined by

$$\rho_0(\varphi) = (\tilde{L}_0\varphi^6 + L_0\varphi^4 + R_0\varphi^2)p_0^{-d} \, ,$$

with $\tilde{L}_0 > 0$, will have a bounded flow for $\vec{v}_h$ only if, given $\tilde{L}_0$, we fix L_0 conveniently so that, by further choosing also the inverse temperature β conveniently, the two parameters $\lambda_0^{(4)}(L_0, \beta), \mu_0(L_0, \beta)$ have the correct value in terms of $\tilde{L}_0$.

In other words, we see that, if $d = 3$ and $\gamma = 0$, one does not expect to see a *normal* or *trivial* critical behavior with two-point functions decaying proportionally to $|x - y|^{-1}$, i.e., proportionally to the free propagator — *unless* the model has one more free parameter to adjust besides the temperature, and the parameter should be such that, by changing it, one affects the ratio $\tilde{L}_0/L_0$ of the nonquadratic terms in the a priorihigh-temperature distribution.

Clearly we do not have this freedom in the model of chapter 2, as $\tilde{L}_0 \equiv 0$ there; and there is no reason, therefore, to expect normal critical behavior in the model of chapter 2 or, more generally, in generic spin models with a prioridistribution with no free parameters.

In a pure φ^4-model with no free parameter $\tilde{L}_0$, what we see is that (12.1) implies that $\lambda_{-1}^{(6)} \neq 0$ and, then, at best stays bounded. This implies, however, divergences in the recursive construction of α_h, even if $\lambda_h^{(4)}$, μ_h behaved well (which they do not, by (12.1), either). In other words, we find ourselves in a situation in which, in the previous sections, we introduced the anomalous dimension.

The interpretation is that α_h (and $\lambda_h^{(4)}$) grow because what is really happening is that the effective potential becomes of longer and longer range because we are looking at it on a wrong scale, and, eventually, we cannot any longer follow it by perturbation methods.

The introduction of the more flexible anomalous scaling procedure to define the effective potentials solves the problem of the evolution of α_h, which disappears from the scene being *replaced* by the evolution of the Z_h. But this *does not* solve the problem of the divergence of $\lambda_h^{(4)}$, which is left, as in (12.1), a relevant operator (at least if we want to think that $\vec{v}_h = (\lambda_h^{(6)}, \lambda_h^{(4)}, \mu_h)$ are small, to apply perturbation theory).

In other words, it is not possible to keep the running couplings flow bounded by simply allowing for an anomalous dimension, at least if one keeps hoping to treat the problem perturbatively, i.e., to use effective potentials and couplings determined by a perturbatively defined beta functional.

Chapter 12

The only hope one has in order to have a bounded flow that can be studied by perturbation theory is that the anomalous flow, which can be easily evaluated to second order (in terms of suitable, numerically computable, constants $\bar{\beta}, \bar{\beta}', \bar{\beta}'', \ldots$) as

$$\lambda^{(6)}_{h-1} = \left(\frac{Z_h}{Z_{h-1}}\right)^3 \left(\lambda^{(6)}_h - \bar{\beta}\lambda^{(6)2}_h - \bar{\beta}'\lambda^{(4)2}_h + \ldots\right),$$

$$\lambda^{(4)}_{h-1} = 2\left(\frac{Z_h}{Z_{h-1}}\right)^2 \left(\lambda^{(4)}_h + \ldots\right),$$

$$\mu_{h-1} = 4\left(\frac{Z_h}{Z_{h-1}}\right)\left(\mu_h + \ldots\right),$$

$$1 = \left(\frac{Z_h}{Z_{h-1}}\right)\left(1 + \bar{\beta}''\lambda^{(6)2}_h + \bar{\beta}'''\lambda^{(4)2}_h + \ldots\right),$$

(12.3)

has a fixed point with $Z_h/Z_{h-1} = 2^{-\eta}$ and $\vec{w} = (\lambda^{(6)}, \lambda^{(4)}, \mu)$ such that

$$\lambda^{(6)} = 2^{-3\eta}\left(\lambda^{(6)} - \bar{\beta}\lambda^{(6)2} - \bar{\beta}'\lambda^{(4)2} + \ldots\right),$$

$$\lambda^{(4)} = 2\,2^{-2\eta}\left(\lambda^{(4)} + \ldots\right),$$

$$\mu = 4\,2^{-\eta}\left(\mu + \bar{\beta}''\lambda^{(6)2} + \ldots\right),$$

$$1 = 2^{-\eta}\left(1 + \bar{\beta}'''\lambda^{(6)2} + \ldots\right),$$

(12.4)

or $\vec{w} = B(\vec{w})$, with η also defined by (12.4).

And, furthermore, the fixed point $\vec{w}$ happens to be small enough to think it *reasonable* that $B(\vec{w})$ evaluated to second order is a good approximation to $B(\vec{w})$ itself (assumed existent; which is not obvious because of the remarks at the end of chapter 10). The latter property depends on the relative size of $\bar{\beta}', \bar{\beta}, \bar{\beta}''$, i.e., on the numerical value of $\bar{\beta}''/\bar{\beta}, \bar{\beta}'/\bar{\beta}$ that should be *small*.

In this way we would have the possibility of thinking that, if the scaling parameter η is appropriate, the running couplings evolve by staying small, so that the critical point is described by a pair Schwinger function decaying differently from the free propagator (by the heuristic arguments of chapter 10), i.e., not proportionally to $|x - y|^{-1}$ (but as $|x - y|^{-1-\eta}$).

The dimensionless effective potentials, computed with the correct scaling (determined by the solutions η, v of (12.4)), can be regarded as a short-range potential that stays small on all scales, converging to a nonzero limit as $h \to -\infty$.

Of course one could try to look for a nontrivial fixed point directly for the *normal scaling* recursion (12.1). But in this case the linear terms in (12.1) for $\lambda^{(6)}$ and for α drop out of the fixed point equation, and the equation will necessarily involve the third-order terms, and one would have even more doubts about the use of the perturbation theory leading to (12.1) itself.

The only known attempt to a more detailed analysis of the problem is the following ([WF]). The coefficients $\bar{\beta}, \bar{\beta}', \bar{\beta}, \ldots$ of the beta function are expressed as integrals involving products of propagators depending on the space dimension d. If the d dependence is taken into account, the (12.1) or (12.4) takes a more general form. For instance, (12.4) becomes (again writing only a few selected second order terms)

$$\lambda_{h-1}^{(6)} = 2^{6-2d}(\frac{Z_h}{Z_{h-1}})^3(\lambda_h^{(6)} - \bar{\beta}(d)\lambda_h^{(6)2} - \bar{\beta}(d)'\lambda_h^{(4)2} + \ldots) \, ,$$

$$\lambda_{h-1}^{(4)} = 2^{4-d}(\frac{Z_h}{Z_{h-1}})^2(\lambda_h^{(4)} + \ldots) \, ,$$

$$\mu_{h-1} = 4(\frac{Z_h}{Z_{h-1}})(\mu_h + \ldots) \, ,$$

$$1 = (\frac{Z_h}{Z_{h-1}})(1 + \bar{\beta}(d)''\lambda_h^{(6)2} + \bar{\beta}(d)'''\lambda_h^{(4)2} + \ldots) \, ,$$

$$(12.5)$$

and (12.1) is similarly changed.

The coefficients $\bar{\beta} \cdot (d)$ turn out to depend on d in a way that admits an analytic interpolation between $d = 3$ and $d = 4$. Of course there is *a considerable ambiguity* in determining the interpolation; even the linear terms are really determined for integer d, and to say that they are as in (12.5) is only one possibility (in some sense quite natural but, nevertheless, arbitrary).

One chooses one particular interpolation (*reasonable* or *natural*) and then one can regard (12.5) as a one-parameter family of recursion relations. A similar procedure could be followed starting from (12.1) (and one would find (12.5) with other coefficients and $Z_h/Z_{h-1} \equiv 1$, plus an equation for α_h replacing the last of (12.1)).

One finds, unless one chooses the analytic interpolation of the $\bar{\beta}$·'s in a *strange way*, that for $d = 4 - \varepsilon$ the anomalous equations (12.5) *do admit* a nontrivial fixed point ($\lambda^{(4)} = O(\varepsilon), \lambda^{(6)} = O(\varepsilon^2) = \mu, Z_h/Z_{h-1} = 2^\eta, \eta = O(\varepsilon^2)$), which can be computed as a formal power series in ε ([WF]). The first nontrivial term in such expansion can be evaluated at $\varepsilon = 1$ and it is still *somewhat small* at $\varepsilon = 1$. So that one is led to conjecture that the full anomalous recursion $\vec{w}' = B(\vec{w})$ has a nontrivial fixed point if $d = 3$, with η computed as above to first nontrivial order in ε (but $\varepsilon = 1!$).

If one tried to do the same analysis with normal scaling, one would not be able to find a nontrivial solution even for ε small, because the second-order equations for α or $\lambda^{(4)}$ do not contain α^2, and this implies that a nontrivial solution would have $O(1)$ values for the couplings even for ε small; hence one cannot really consider that, in such situations, the beta function is given with good approximation by a power series expansion in the couplings truncated to second order or to a finite order.

We thus see that if $d = 3, \gamma = 0$ we expect, for generic a priorispin distribution, a critical point with anomalous dimension.

Unfortunately, the above analysis rests on a somewhat too conjectural basis and it would be desirable to have a stronger evidence for such a nontrivial anomalous fixed point.

For this reason, starting with Wilson ([W1], [W2], [W3], [W4], models that are simpler than the one of chapter 2 have been introduced and investigated ([G1], §22; [G2]; [Rm1]). These are models so simplified that the beta function can be given a well-defined meaning even beyond perturbation theory. They can therefore be used to test the above ideas and methods: for instance, the existence of a nontrivial fixed point was found, numerically, by Wilson ([W2], [W3]), in one such model.

Only recently has it become possible to establish rigorously the existence of a nontrivial fixed point for some of the so-called *hierarchical models* in dimension $d = 3$ (see [KW1], [KW2]) — a remarkable achievement of the new techniques developed to construct computer-assisted proofs.

We can say that, to a large extent, the renormalization group only rarely can be carried to a high level of mathematical rigor: with some remarkable exceptions (see [G1] for a list of references). It does, however, provide a beautiful and rich general scheme to set problems in perspective, gaining some deep insights that go well beyond the ones provided by the previously known theoretical approaches.

Chapter 13
Bose Condensation: Reformulation

This problem will be studied at an heuristic level, i.e., in a way similar to the theory of the critical point in $d = 4 - \varepsilon$ dimensions (see chapters 9 and 10). One should not forget that the latter problem, considered by many one of the major successes of the renormalization group approach, is still awaiting a rigorous formulation, *not only for $\varepsilon = 1$, but also for any $\varepsilon > 0$*. Therefore, we think that a formal theory is nevertheless an interesting way to attack the Bose condensation problem in $d = 3$ dimensions. This will be, in fact, the best illustration of the methods we try to illustrate about the renormalization group, as it embodies all the ideas discussed in the various cases met so far. The analysis that follows is due to [B], and we follow his treatment with some minor changes.

What follows could also be regarded as a test of the usual claim that the zero temperature Bose gas has a linear dispersion relation for small momenta. This property is considered typical of *superfluid behavior* and was first verified by Bogoliubov ([Bo]) in an approximate exactly soluble model, the so-called *Bogoliubov model*. After that, the superfluid behavior hypothesis received strong support from rough perturbative arguments (see, for example, [ADG], where it is possible to find relevant references). More recently, more convincing arguments were presented in the papers of [NN] and [PS].

The problem is formulated as a functional integration in chapter 3. The functional integral is

$$\frac{\int \varphi_{x_1}^{\sigma_1} \cdots \varphi_{x_n}^{\sigma_n} \, e^{-V_\Lambda(\varphi)} \, P(d\varphi)}{\int e^{-V_\Lambda(\varphi)} \, P(d\varphi)} , \tag{13.1}$$

with $\Lambda = [-\frac{1}{2}\beta, \frac{1}{2}\beta] \times [-\frac{1}{2}L, \frac{1}{2}L]^3$,

$$V_\Lambda(\varphi) = \lambda \int v(\vec{x} - \vec{y})\delta(x^0 - y^0) \, (\varphi_x^+ \varphi_x^-)(\varphi_y^+ \varphi_y^-) \, dx \, dy + \\ + \nu \int_\Lambda \varphi_x^+ \varphi_x^- \, dx , \tag{13.2}$$

and $P(d\varphi)$ is the Gaussian measure with propagator $\langle \varphi_x^\sigma \varphi_y^\sigma \rangle = 0$ and

$$\langle \varphi_x^- \varphi_y^+ \rangle \equiv S(x) = \rho + \frac{1}{(2\pi)^4} \int dk \frac{e^{-ikx}}{-ik_0 + \vec{k}^2/2m} . \tag{13.3}$$

As explained in chapter 3, the problem is to fix $\rho > 0$ and to show that one can fix $\nu = \nu(\lambda, \rho)$ so that the interacting pair Schwinger function converges to ρ as $x \to \infty$. This is interpreted (see chapter 3) by saying that the Schwinger functions thus obtained describe a Bose condensed state with condensate density ρ equal to the off-diagonal long-range order parameter.

We also, and mainly, want to understand the behavior of $S(x) - \rho$, and in particular we would like to check the prediction (following from superfluid behavior hypothesis) that for large x the function $S(x) - \rho$ vanishes at ∞ as some inverse power of $v^2 x_0^2 + \vec{x}^2$, where v is the *sound velocity* (see [ADG]). Note that, in the Bogoliubov model, $v^2 = v_B^2 = 2m^{-1}\lambda \hat{v}(\vec{0})\rho$, if $\hat{v}(\vec{0}) \equiv \int v(\vec{x})d^3\vec{x}$.

We begin by supposing that the problem has an ultraviolet cutoff on the scale p_0 of the range of the potential. This means that we consider the functional integral (13.1) with a field φ with propagator $\langle \varphi_x^\sigma \varphi_y^\sigma \rangle = 0$ and

$$g_{\leq 0}(x - y) = \langle \varphi_x^- \varphi_y^+ \rangle = \rho + \frac{1}{(2\pi)^4} \int dk \frac{e^{-ik(x-y)} t_0(k)}{-ik_0 + \vec{k}^2/2m} , \qquad (13.4)$$

where $t_0(k)$ is a smooth ultraviolet cutoff on scale p_0, which we choose to be a regularization of the characteristic function of the set $\{f(k) \equiv (k_0^2 + \frac{\vec{k}^2}{2m}\frac{p_0^2}{2m}) \leq (\frac{p_0^2}{2m})^2\}$. It will be a function $t_0(k) = 0$ if $f(k) \geq 1$ and $t(k) = 1$ if $f(k) \leq 1/\gamma$.

The assumed presence of the ultraviolet cutoff on the scale of the interaction potential is reasonable only if $\rho p_0^{-3} \ll 1$, i.e., only if there is, in mean, less than one particle in a cube with the side equal to the range of the potential (which we call p_0^{-1}).

If such a situation is realized, the interaction potential can be replaced by the simpler

$$V_\Lambda^{local}(\varphi) = \lambda \hat{v}(\vec{0}) \int_\Lambda (\varphi_x^+ \varphi_x^-)^2 \, dx + \nu \int_\Lambda \varphi_x^+ \varphi_x^- \, dx . \qquad (13.5)$$

Therefore we shall study the functional integral

$$e^{-V_\Lambda^{local}(\varphi)} P(d\varphi) , \qquad (13.6)$$

with $P(d\varphi)$ being the Gaussian measure with propagator (13.4).

The form of the propagator (13.4) shows that the field $\varphi_x^\pm$ can be represented as

$$\varphi_x^\pm = \xi^\pm + \psi_x^\pm , \qquad (13.7)$$

where ξ, ψ are independent, with propagators $\langle \xi^\sigma \xi^\sigma \rangle = 0$, $\langle \varphi_x^\sigma \varphi_y^\sigma \rangle = 0$, $\langle \psi_x^- \psi_y^+ \rangle$ given by the integral in (13.4) and $\langle \xi^+ \xi^- \rangle = \rho$.

If one defines $W^{(-\infty)}(\xi) = -\frac{1}{\Lambda} \log \int e^{-V_\Lambda(\xi+\psi)} P(d\psi)$, we see that the computation of $\langle \xi^+ \xi^- \rangle$ in the presence of interaction will lead (if we set $\xi^\pm = \xi_1 \pm i\xi_2$) to the integral

$$\rho = \int \frac{d\xi_1 d\xi_2}{2\pi\rho} (\xi_1^2 + \xi_2^2) \, e^{-(\xi_1^2 + \xi_2^2)/\rho} \, e^{-(W^\infty(\xi) - \bar{W})|\Lambda|} , \qquad (13.8)$$

where $\bar{W}$ is a normalization constant, and the equality to ρ of the above integral is just the requirement that the condensate density should be ρ. Therefore, equality (13.8) can hold if and only if the function $W^\infty(\xi)$, which is a function of the product $\xi^+ \xi^-$, by symmetry considerations, reaches its minimum at $\xi^+ \xi^- = \rho$. And in this case $\xi^+ \xi^-$ will be a sure random variable, provided the minimum is nondegenerate. The only rigorous general result in this direction is provided by the work in [Gi].

Hence ν, i.e., the chemical potential, is simply determined by imposing the condition $\partial_\rho W^\infty(\rho) = 0$; and we expect that if ν is so fixed, then the Schwinger functions will be computable by simply setting $\xi^+ = \xi^- = \sqrt{\rho}$ in the functional integral $e^{-V_\rho(\psi)} P(d\psi)$ with

$$V_\rho(\psi) =$$
$$= \lambda\hat{v}(0) \int_\Lambda (\psi_x^+ \psi_x^-)^2 \, dx + 2\lambda\hat{v}(0)\sqrt{\rho} \int_\Lambda \psi_x^+ \psi_x^- (\psi_x^+ + \psi_x^-) \, dx +$$
$$+ (4\lambda\hat{v}(0)\rho + \nu) \int_\Lambda \psi_x^+ \psi_x^- \, dx + \lambda\hat{v}(0)\rho \int_\Lambda ((\psi_x^+)^2 + (\psi_x^-)^2) \, dx +$$
$$+ |\Lambda|(\nu\rho + \lambda\hat{v}(0)\rho^2) , \qquad (13.9)$$

where $\langle \psi_x^\sigma \psi_y^\sigma \rangle = 0$, $\langle \psi_x^- \psi_y^+ \rangle$ is given by the integral in (13.4); this model is called the *Bogoliubov approximation*. One has to show that, if ν is suitably fixed, then $-\log \int \exp -V_\rho(\psi) P(d\psi)$ has a derivative with respect to ρ at fixed ν vanishing (and, in fact, corresponding to a minimum).

The well-known analysis of Hughenoltz and Pines ([HP]), based on formal perturbation theory, shows that this condition for ν is formally equivalent to the following one

$$\Sigma_{-+}(0) = \Sigma_{++}(0) , \qquad (13.10)$$

where $\Sigma_{\sigma_1 \sigma_2}(k)$ is the Fourier transform of the sum of all one-particle irreducible graphs (connected graphs that cannot become disconnected by cutting one leg) with two external lines $\psi_x^{\sigma_1}$, $\psi_y^{\sigma_2}$; see [ADG].

The Bogoliubov model is obtained by neglecting in (13.9) the quartic and cubic terms. It is easy to see that, in this approximation, the condition (13.10) should be $2\lambda\hat{v}(0)\rho + \nu = 0$.

Therefore the functional integral (13.6) should be written as

$$e^{-V^{(0)}(\psi)} P_B(d\psi) , \tag{13.11}$$

where $P_B(d\psi)$ is the Gaussian measure obtained by including the quadratic terms in (13.9) into the free measure

$$P_B(d\psi) \equiv P(d\psi) e^{-\lambda\hat{v}(0)\rho \int_\Lambda (\psi_x^+ + \psi_x^-)^2)dx} , \tag{13.12}$$

and ν is replaced by $-2\lambda\hat{v}(0)\rho + \nu^0$, where ν^0 is the correction to the chemical potential due to the quartic and cubic interaction terms, so that

$$V^{(0)}(\psi) = \lambda\hat{v}(0)\int_\Lambda (\psi_x^+ \psi_x^-)^2 \, dx +$$
$$+ 2\lambda\hat{v}(0)\sqrt{\rho}\int_\Lambda \psi_x^+ \psi_x^- (\psi_x^+ + \psi_x^-) \, dx + \nu^0 \int_\Lambda \psi_x^+ \psi_x^- \, dx , \tag{13.13}$$

and the parameter ν^0 has to be determined, as above, so that condition (13.10) is satisfied.

It is convenient, before proceeding, to change the basic fields and to perform a rescaling (amounting at fixing $\rho = 1$); we set

$$\chi_x^\pm = \frac{1}{\sqrt{2\rho}}(\psi^+ \pm \psi^-), \qquad \psi^\pm = \sqrt{\frac{\rho}{2}}(\chi^+ \pm \chi^-) ,$$
$$\varepsilon = \lambda\hat{v}(\vec{0})\rho\frac{2m}{p_0^2} , \tag{13.14}$$

so that $V^{(0)}(\chi)$ becomes a function of χ

$$V^{(0)}(\chi) = \frac{p_0^2\rho}{2m}\left(\frac{\varepsilon}{4}\int_\Lambda ((\chi_x^-)^4 - 2(\chi_x^-)^2(\chi_x^+)^2 + (\chi_x^+)^4)dx +\right.$$
$$+ \varepsilon\sqrt{2}\int_\Lambda ((\chi_x^+)^3 - (\chi_x^-)^2\chi_x^+) \, dx + \tag{13.15}$$
$$\left. + \frac{\nu^0}{2}\frac{2m}{p_0^2}\int_\Lambda ((\chi_x^+)^2 - (\chi_x^-)^2) \, dx\right) ,$$

and the propagator of the fields $\chi^\pm$ in the distribution (13.12) is

$$\langle \chi_x^{\sigma_1} \chi_y^{\sigma_2} \rangle = g^{\sigma_1 \sigma_2}(x - y) , \tag{13.16}$$

which, in the limit as $\Lambda \to \infty$, is easily computed

$$g^{\sigma_1 \sigma_2}(x) = \frac{1}{(2\pi)^4} \int e^{-ikx} t_0(k) G_0^{-1}(k)_{\sigma_1 \sigma_2} , \qquad (13.17)$$

where the matrix $G_0(k)$, which we call the *propagator matrix* of (13.12), is defined by

$$G_0(k) = \rho \begin{pmatrix} \frac{\vec{k}^2}{2m} + 4\varepsilon \frac{p_0^2}{2m} t_0(k) & ik_0 \\ -ik_0 & -\frac{\vec{k}^2}{2m} \end{pmatrix} , \qquad (13.18)$$

where the first row and column correspond to $\sigma = +$ and the second row and column correspond to $\sigma = -$, so that $G_0(k)_{++}$ is the first element of the matrix. In appendix 5 we discuss the derivation of (13.18) in a form that will be repeatedly used in the following.

The ψ fields Schwinger functions in the Bogoliubov model are, therefore, immediately deduced from (13.17). The main feature of the approximation is that, if it is assumed, it follows that the singularity at $k = 0$ of the propagator is determined by the determinant of the matrix $G_0(k)$ (because the propagator is just the inverse of G_0). And it is clear that the determinant vanishes as $k_0^2 + 4\varepsilon \frac{p_0^2}{2m} \frac{\vec{k}^2}{2m}$, so that the *speed of sound* $c(\vec{k})$ on scales $|\vec{k}| \ll p_0$ is, if ε is defined as in (13.14),

$$c(\vec{k})^2 = \varepsilon v_0^2, \qquad v_0 = \frac{p_0}{m} , \qquad (13.19)$$

instead of $\frac{\vec{k}^2}{4m^2} \xrightarrow[\vec{k} \to 0]{} 0$, which would be the free field result.

Checking the validity of the superfluid behavior hypothesis means, mainly, to check that the anomalous (with respect to the free case) behavior (13.19) remains valid when the potential $V^{(0)}$ in (13.11), (13.13) is not set equal to 0, possibly with a different value of $c(\vec{k})$.

In the next chapter we start the check by constructing an algorithm to evaluate recursively the integral (13.11), defining effective potentials $V^{(h)}$ on scales $\gamma^h p_0$, where γ is a scaling parameter.

In passing from $V^{(h+1)}$ to $V^{(h)}$ we shall see that the effective potential retains, essentially, the same form (13.13) with new coefficients that approach 0 as $h \to -\infty$ *provided* the Gaussian integral P_B is replaced by a new Gaussian integral which, besides having a cutoff function $t_h(k) = t_0(\gamma^{-h} k)$ (i.e., a lower ultraviolet cutoff), also has a new sound speed c_h, i.e., a new ratio between the coefficients of $\vec{k}^2$ and k_0^2 in the propagator $t_h(k) G_h(k)^{-1}$ (that is, in $\det G_h(k)$).

Hence, there is superfluid behavior, if c_h has a finite limit as $h \to -\infty$.

In particular, the analysis will imply that the pair Schwinger function $S(x)$ tends to ρ as $x \to \infty$, i.e., the existence of long-range order and, therefore, an important part of the usual picture of the Bose condensation.

Our analysis will be performed "to leading order" in the running couplings; we shall show also that the results *do not change* if the analysis is pushed to an arbitrarily high order in the running couplings, as long as it is finite. The leading order turns out to coincide with the "one-loop approximation" in the beta function and in the running couplings. Going beyond all orders of perturbation theory requires a nonperturbative treatment of the "large fields", which is hard in this case because the fields are complex valued.

Chapter 14
Bose Condensation: Effective Potentials

We begin by defining the running couplings: once their flow, i.e., their scale dependence, is controlled, one can straightforwardly control the Schwinger functions by proceeding as in chapter 10.

Given $V^{(0)}(\chi)$ as in (13.15) we set

$$\chi = \chi^{(\leq -1)} + \chi^{(0)} , \tag{14.1}$$

where $\chi^{(0)}$ and $\chi^{(\leq -1)}$ have propagators

$$g_0^{\sigma_1 \sigma_2}(k) = T_0(k) G_0^{-1}(k)_{\sigma_1 \sigma_2}, \qquad \tilde{g}_{\leq -1}^{\sigma_1 \sigma_2}(k) = t_{-1}(k) G_0^{-1}(k)_{\sigma_1 \sigma_2} ,$$
$$T_0(k) = t_0(k) - t_0(\gamma k), \qquad t_{-1}(k) = t_0(\gamma k) , \tag{14.2}$$

which decompose the propagator $g_{\leq 0}$ in (13.17) if the matrix $G_0(k)$ is defined as in (13.18). And $\gamma > 1$ is a scaling parameter.

The integral for the "partition function" can, therefore, be written

$$I = \int e^{-V^{(0)}(\chi^{(\leq -1)} + \chi^{(0)})} P_0(d\chi^{(0)}) \tilde{P}_{\leq -1}(d\chi^{(\leq -1)}) . \tag{14.3}$$

After the integration over $\chi^{(0)}$ equation (14.3) becomes

$$I = \int e^{-\tilde{V}^{(-1)}(\chi^{(\leq -1)})} \tilde{P}_{\leq -1}(d\chi^{(\leq -1)}) , \tag{14.4}$$

where $\chi^{(\leq -1)}$ has propagator $\tilde{g}_{\leq -1}$ and $\chi^{(0)}$ has propagator g_0.

Since we suspect an anomalous behavior, we look at the quadratic part of $\tilde{V}^{(-1)}$; as in chapter 10, before splitting $\chi^{(\leq -1)}$ into $\chi^{(-1)} + \chi^{(\leq -2)}$ and iterating the integrations, we shall split off the quadratic terms in $\tilde{V}^{(-1)}$ the marginal ones, and include them in the $\tilde{P}$ measure, thus changing it to a measure that we can denote $P_{\leq -1}(d\chi)$ with a new propagator matrix $t_{-1}(k)^{-1} G_{-1}^{-1}(k)$. Once G_{-1} is determined we shall split its *inverse* (c.f.r. chapter 10) as $T_{-1}(k) G_{-1}^{-1}(k) + t_{-2}(k) G_{-1}^{-1}(k)$, with $T_{-1}(k) \equiv t_{-1}(k) - t_{-2}(k)$ and $t_{-2}(k) = t_{-1}(\gamma k)$. We generate in this way a decomposition of the field $\chi^{(\leq -1)} = \chi^{(-1)} + \chi^{(\leq -2)}$. This will allow us to define $\tilde{V}^{(-2)}$ and to proceed recursively.

Calling $V^{(-1)}$ the part of $\tilde{V}^{(-1)}$ without the marginal quadratic operators, we establish the chain of identities

$$
\begin{aligned}
I &= \int e^{-V^{(0)}(x^{(\leq -1)}+x^{(0)})} P_0(d\chi^{(0)}) \tilde{P}_{\leq -1}(d\chi^{(\leq -1)}) = \\
&= \int e^{-\tilde{V}^{(-1)}(x^{(\leq -1)})} \tilde{P}_{\leq -1}(d\chi^{(\leq -1)}) = \dots \\
\dots &= \int e^{-\tilde{V}^{(h)}(x^{(\leq h)})} \tilde{P}_{\leq h}(d\chi^{(\leq h)}) = \\
&= \int e^{-V^{(h)}(x^{(\leq h-1)}+x^{(h)})} P_h(d\chi^{(h)}) \tilde{P}_{\leq h-1}(d\chi^{(\leq h-1)}) .
\end{aligned}
\tag{14.5}
$$

If the quadratic part of $\tilde{V}^{(h)}(\chi^{(\leq h)})$ that is marginal is described by a matrix $\Delta_h^{\sigma\sigma'}(k)$, i.e., if the marginal quadratic part of $\tilde{V}^{(\leq h)}$ has the form ($\chi_x^{\sigma} = (2\pi)^{-2} \int dk \exp(i\sigma kx)\chi_k^{\sigma}$),

$$
\sum_{\sigma,\sigma'=\pm} \int \Delta_h^{\sigma\sigma'}(k)\chi_{\sigma k}^{\sigma}\chi_{-\sigma' k}^{\sigma'} \, dk ,
\tag{14.6}
$$

then given $G_{h+1}(k)$, the propagator matrix $G_h(k)$ will, naturally, be defined by

$$
\frac{1}{2} t_h(k)^{-1} G_{h+1}(k) + \Delta_h \equiv \frac{1}{2} t_h(k)^{-1} G_h(k) .
\tag{14.7}
$$

Therefore the "only task" in order to define completely the recursive algorithm for the computation of the partition function integral I via (14.5) is the identification of the relevant and marginal operators in $\tilde{V}^{(h)}(\chi)$.

Without repeating the heuristic analysis leading to the identification of the relevant and marginal terms, we simply define the relevant and marginal operators by a localization operator $\mathcal{L}$, as in all the previously treated cases. The reason this is a natural definition (and, up to trivialities, the only possible one) should be clear to the reader who followed us in the preceding chapters. The reason it is interesting is that it leads to an understanding of superfluid behavior.

It will turn out that the appropriate representation of the fields $\chi^{(\leq h)\pm}$ in terms of dimensionless fields (if one wanted to proceed as in chapter 6) is

$$
\chi_x^- = \gamma^h \overline{\chi}_{\gamma^h x}^-, \qquad \chi_x^+ = \gamma^{2h} \overline{\chi}_{\gamma^h x}^+ ,
\tag{14.8}
$$

where the different scaling of χ^+ and χ^- is a new essential feature of the problem we are discussing. Accepting the above scaling, it is clear that the marginal operators are

$$F_{40} = \int_\Lambda (\chi_x^-)^4 \, dx, \quad F_{21} = \int_\Lambda (\chi_x^-)^2(\chi_x^+) \, dx \ ,$$

$$F_{02} = \int_\Lambda (\chi_x^+)^2 \, dx \ ,$$

$$D_{tt} = -(\frac{2m}{p_0^2})^2 \int_\Lambda (\partial_{x_0}\chi_x^-)^2 \, dx \ , \tag{14.9}$$

$$D_{ss} = -p_0^{-2} \int_\Lambda (\partial_{\bar{x}}\chi_x^-)^2 \, dx, \quad D_t = -\frac{2m}{p_0^2} \int_\Lambda \chi_x^+ \partial_{x_0}\chi_x^- \, dx \ .$$

We did not include in the list the operator $-\int_\Lambda dx \chi_x^+ \partial_{\bar{x}}\chi_x^- = (1/\Lambda)\sum_k \chi_k^+ i\vec{k}\chi_{\bar{k}}^-$, since it is identically zero; in fact, the fields χ_k^σ are even functions of $\vec{k}$.[1]

Let us now consider the relevant operators. As usual, the operators $\int_\Lambda dx \chi_x^- \partial \chi_x^-$ and $\int_\Lambda dx (\chi_x^-)^2 \partial \chi_x^-$ are absent, since they are integrals of total derivatives and the fields satisfy periodic boundary conditions. The operator

$$F_{20} = \int_\Lambda (\chi_x^-)^2 dx \tag{14.10}$$

is the only relevant operator left. In fact the operators of degree 1, which would also be relevant, vanish because the fields $\psi^\pm$, hence the $\chi^\pm$, have zero average.[2] Also, the operators $\int dx \chi_x^+ \chi_x^-$ and $\int dx (\chi_x^-)^3$, which would also be relevant, are absent, as a consequence of the particular structure of the potential (13.9), which implies that the local monomials of order 2 and 3 in the fields $\psi^\pm$ *must* appear in the following combina-

---

[1] Here periodic boundary conditions are important; as noted in chapter 2 this, however, implies a discussion somewhat more detailed, which we skip (as already done in the previous cases).

[2] This holds essentially by construction; note, however, that the construction is possible because the free Bose gas has condensation for $d = 3$. This would also be possible if $d = 2$ because our reference measure is the Bogoliubov distribution rather than the free Bose gas, and the Bogoliubov propagator has a singularity like $k_0^2 + \varepsilon v_0^2 \bar{k}^2$ rather than $k_0^2 + (\bar{k}^2/2m)^2$. But it would definitely be impossible if $d = 1$. This makes us think, in principle, of adapting the methods discussed here to the $d = 2$ case. But the work remains to be done.

tions

$$(\psi_x^+)^3 + (\psi_x^-)^3 = 2\frac{\rho^{3/2}}{2^{3/2}}((\chi_x^+)^3 + 3\chi^+(\chi^-)^2) \,,$$

$$\psi_x^+ \psi_x^- (\psi_x^+ + \psi_x^-) = 2\frac{\rho^{3/2}}{2^{3/2}}((\chi_x^+)^3 - (\chi_x^+)(\chi_x^+)^2) \,,$$

$$\psi_x^+ \psi_x^- = \frac{\rho}{2}((\chi_x^+)^2 - (\chi_x^-)^2) \,,$$

$$(\psi_x^+)^2 + (\psi_x^-)^2 = 2\frac{\rho}{2}((\chi_x^+)^2 + (\chi_x^-)^2) \,.$$

$$(14.11)$$

Therefore the localization operator $\mathcal{L}$ that we introduce is identical with the first of (8.2) for the monomials of degree 4 in χ^-, and it is similarly defined for all the other monomials, giving rise to local marginal operators, while for the monomial $\chi_x^- \chi_y^-$ it will be defined by (8.3).[3] Finally, we have to consider the localization of the monomial $\chi_x^+ \chi_y^-$, which has dimension 1; its local part is obtained by a first-order Taylor expansion of the field χ_y^- and, as explained before, the term of order zero should not give any contribution.

Therefore, the result of the application of the localization operator to $\tilde{V}^{(h)}(\chi)$ will be, necessarily,

$$\mathcal{L}\tilde{V}^{(h)}(\chi) = \frac{p_0^2 \rho}{2m}\left(\lambda_h F_{40} + \mu_h F_{21} - \gamma^{2h}\nu_h F_{20} + \right.$$
$$\left. + 2z_h F_{02} + 2\zeta_h D_{tt} + 2\alpha_h D_{ss} + 2d_h D_t\right) \,.$$

$$(14.12)$$

The *running couplings* λ_h, μ_h, ν_h and the *renormalization constants* z_h, ζ_h, α_h, d_h are defined by (14.12); the factors 2 are introduced only for consistency with the definitions used in [B]. The dimension fixing factor $\frac{p_0^2 \rho}{2m}$ is introduced to keep track of the dimensions of the various quantities (its physical dimension is that of an action density, in space time; recall also that[4] we set, once and for all, $\hbar = 1$).

Comparing (14.12) with (13.15), we see that $V^{(0)}$ contains some irrelevant operators as well as a few relevant or marginal ones; the latter appear with coefficients

$$\lambda_0 = \frac{\varepsilon}{4}, \quad \mu_0 = -\varepsilon\sqrt{2}, \quad \nu_0 = \frac{\nu^0}{2}\frac{2m}{p_0^2}, \quad z_0 = \nu_0 \,,$$
$$\alpha_0 = \zeta_0 = d_0 = 0 \,.$$

$$(14.13)$$

---

[3] Note that this is the obvious extension to the present case of the ideas used to discuss the previous cases; we feel that repeating the heuristic motivation is useless here.

[4] Unfortunately.

It follows from (14.12) that, in general, the matrix $\Delta_{(h)}$ of the quadratic marginal operators, c.f.r. (14.6), will be

$$\Delta_{(h)} = \rho \begin{pmatrix} 2z_h \frac{p_0^2}{2m} & id_h k_0 \\ -id_h k_0 & -2\frac{2m}{p_0^2} k_0^2 \zeta_h - 2\alpha_h \frac{\vec{k}^2}{2m} \end{pmatrix} . \qquad (14.14)$$

Therefore the recursive scheme to evaluate the integral I can be summarized as follows.

Suppose that it has been proved that

$$I = \int e^{-V^{(h')}(x^{(\leq h'-1)}+x^{(h')})} P_{h'}(\chi^{(h')}) \tilde{P}_{\leq h'-1}(dx^{(\leq h'-1)}) \qquad (14.15)$$

holds for $h' = 0, -1, \ldots, h+1$, with suitably chosen $V^{(h')}$ and with the Gaussian integrals having propagators

$$\begin{aligned} T_{h'} G_{h'}^{-1}, \qquad & t_{h'}(k) G_{h'}^{-1} , \\ t_{h'}(k) = t_0(\gamma^{-h'} k), \qquad & T_{h'}(k) = t_{h'}(k) - t_{h'-1}(k) , \end{aligned} \qquad (14.16)$$

and with $G_{h'}(k)$ given by

$$G_{h'}(k) = \rho \begin{pmatrix} \frac{\vec{k}^2}{2m} + \frac{2p_0^2}{m} Z_{h'} & ik_0 E_{h'} \\ -ik_0 E_{h'} & -\left(\frac{\vec{k}^2}{2m} + \frac{8B_{h'} m k_0^2}{p_0^2} + \frac{2A_{h'} \vec{k}^2}{m}\right) \end{pmatrix} . \qquad (14.17)$$

Then performing the integration over $\chi^{(h+1)}$ we write

$$e^{-\tilde{V}^{(h)}(x^{(\leq h)})} = \int e^{-V^{(h+1)}(x^{(\leq h)}+x^{(h+1)})} P_{h+1}(dx^{(h)}) , \qquad (14.18)$$

and defining $V^{(h)}(\chi) = \tilde{V}^{(h)}(\chi) - \mathcal{L}_2 \tilde{V}^{(h)}(\chi)$, if $\mathcal{L}_2 \tilde{V}^{(h)}$ is the part of $\mathcal{L} \tilde{V}^{(h)}$ containing the three marginal terms of degree 2, we see that

$$I = \int e^{-V^{(h)}(x^{(\leq h-1)}+x^{(h)})} P_h(\chi^{(h)}) \tilde{P}_{\leq h-1}(dx^{(\leq h-1)}) \qquad (14.19)$$

if the propagator matrix $G_{(h)}(k)$ is defined by (14.17), (14.14), with

$$\begin{aligned} Z_h(k) &= Z_{h+1}(k) + t_h(k) z_h , \\ B_h(k) &= B_{h+1}(k) + t_h(k) \zeta_h , \\ A_h(k) &= A_{h+1}(k) + t_h(k) \alpha_h , \\ E_h(k) &= E_{h+1}(k) + 2t_h(k) d_h , \end{aligned} \qquad (14.20)$$

with $Z_0 = \varepsilon t_0(k)$, $E_0 = 1$, and $A_0 = B_0 = 0$.

From now on we shall use a sharp cutoff function $t_0(k)$, equal to 1 if $k_0^2 + \frac{p_0^2}{2m}\frac{\vec{k}^2}{2m} < (\frac{p_0^2}{2m})^2$ and equal to 0 otherwise. We shall see that sometimes one also needs the derivatives of $t_h(k)$; they will be treated as the appropriate delta functions. It seems a well-known fact that the delta functions thus introduced, which sometimes even appear raised to some power, never give divergent contributions: whenever they try, the divergent contributions are several and they *cancel exactly*. We have checked that this is the case in our problem too (see appendix 6). This will simplify the calculations considerably as the functions $Z_h(k), A_h(k), B_h(k)$ become constants Z_h, A_h, B_h for k in the region where t_h does not vanish.

The square of the sound speed c_h on scale $\gamma^h p_0$, if ν_0 can be so chosen that the running couplings $\lambda_h, \mu_h, \nu_h \xrightarrow[h \to -\infty]{} 0$, is simply given by the ratio of the coefficients of $\vec{k}^2$ and k_0^2 in the scale h propagator singularity, i.e., in the determinant $\det G_{(h)}(k)$. This is

$$c_h^2 = \frac{4\frac{p_0^2}{2m}\frac{1}{2m}Z_h\,(1 + 4A_h)}{E_h^2 + 16B_h Z_h} = v_0^2 Z_h \frac{1 + 4A_h}{E_h^2 + 16B_h Z_h} \ . \tag{14.21}$$

It was already noted that $c_0^2 = \varepsilon v_0^2$ if $v_0 = \frac{p_0}{m}$ and $\varepsilon = Z_0 = \lambda \hat{v}(\vec{0})\rho$ $2mp_0^{-2}$.

In chapter 15 we write the recursion relation between $(\lambda_h, \mu_h, \nu_h, A_h, B_h, E_h, Z_h)$ at different values of h and show that, if the free parameter ν_0 is suitably fixed, then λ_h, μ_h, ν_h, Z_h, E_h approach 0 (*asymptotic freedom*) at the same rate $1/|h|$, while

$$A_h \xrightarrow[h \to -\infty]{} A_{-\infty}, \qquad B_h \xrightarrow[h \to -\infty]{} B_{-\infty} \ , \tag{14.22}$$

with $A_{-\infty} = O(\varepsilon)$ and $B_{-\infty} = (16\varepsilon)^{-1}(1 + O(\sqrt{\varepsilon}))$, so that the speed of sound has a well defined limit as $h \to -\infty$, given by

$$c_{-\infty}^2 = v_B^2(1 + O(\sqrt{\varepsilon})) \ . \tag{14.23}$$

Chapter 15
The Beta Function
for the Bose Condensation

The renormalization group flow has been defined recursively and summarized in the discussion between (14.15) and (14.20).

It is possible to see that the condition (13.10), which determines the chemical potential given ρ, is satisfied if $\nu_h \xrightarrow[h \to -\infty]{} 0$. We shall check that the flow is asymptotically free, with the running couplings $\lambda_h, \mu_h, \nu_h \xrightarrow[h \to -\infty]{} 0$, if ν_0 is suitably tuned, and the renormalization constants $Z_h, E_h \xrightarrow[h \to -\infty]{} 0$ at the same rate, while $A_h, B_h \xrightarrow[h \to -\infty]{} const$ (so that the speed of sound tends to a finite limit).

The analysis would be totally inconclusive without the following extra property, which is a deep identity of the type of the Ward identities, and which plays for the Bose gas the same role as the vanishing of the beta function plays in the $d = 1$ Fermi gas. This identity (see [B]) is

$$Z_h = -\frac{1}{\sqrt{2}}\mu_h + \gamma^{2h}\nu_h, \qquad h < 0 , \qquad (15.1)$$

while for $h = 0$ it is $Z_0 = -\frac{\mu_0}{\sqrt{2}}$.[1]

In fact, if such relation did not hold, the renormalization group flow would have been very unstable and the results would depend on very tiny details on the initial data, on which no assumptions could be made as we perform only approximate calculations (neglecting high-order corrections). On the other hand, the a priori validity of (15.1) will stabilize the results, which turn out to be no longer dependent on the initial data for the flow, and they remain the same no matter how many perturbative orders for the beta function we take (once the full leading order is accurately computed).

The identity (15.1) holds for the model that we are studying, (13.5), or even for the nonlocal model (13.2) (which could also be studied with the techniques we are presenting and would lead exactly to the same

---

[1] Because in the expression for $V^{(0)}$ we left out an operator proportional to the marginal operator G_{02} with coefficient ν_0, which is a choice that we modified on the following scales $h < 0$ so that $V^{(h)}$ contains no marginal operators at all.

qualitative results). It is an identity for any choice of λ_0 and ν_0, and it holds as an identity valid to all orders of the formal expansion of both sides in powers of λ_0, ν_0.

We reproduce here the proof of (15.1), from [B]. Note that the constants $\lambda_h, \mu_h, Z_h, A_h, B_h, E_h$ can be defined in another equivalent way by the following steps.

1. We first integrate in a single step the fluctuations up to scale $h + 1$, without any free measure renormalization and by using the original representation (13.5) (or, if one dislikes the local model, (13.2)) of the potential. This simply means that we write the propagator (13.4), without the constant term, as $g_{\leq h}(x) + g_{>h}(x)$, where $g_{\leq h}(x)$ and $g_{>h}(x)$ are defined respectively by replacing $t_0(k)$ in (13.4) by $t_0(\gamma^{-h}k)$ and $(t_0(k) - t_0(\gamma^{-h}k))$. The result of the integration defines an effective potential $\tilde{U}_h(\varphi)$ such that

$$\int P(d\varphi)e^{-V_\Lambda(\varphi)} = \int P^{(\leq h)}(d\varphi)e^{-\tilde{U}_h(\varphi)} , \qquad (15.2)$$

where $P^{(\leq h)}(d\varphi)$ has propagator $g_{\leq h}(x)$.

The potential $\tilde{U}_h$ can be written in the general form

$$\tilde{U}_h(\varphi) = \sum_{s=1}^{\infty} \int d\vec{x}d\vec{y} W_{h,s}(\vec{x}, \vec{y})\varphi_{x_1}^+ \cdots \varphi_{x_s}^+ \varphi_{y_1}^- \cdots \varphi_{y_s}^- , \qquad (15.3)$$

where $\vec{x} = (x_1, \ldots, x_s), \vec{y} = (y_1, \ldots, y_s)$.

2. We insert the representation (13.7) of $\varphi^\pm$ in (15.3) collecting the operators containing the monomials $(\prod_{i=1}^{r_1} \psi_{x_i}^+)(\prod_{j=1}^{r_2} \psi_{y_j}^-)$, with $2 \leq r_1 + r_2 \leq 4$, and we localize them, by a Taylor expansion of the fields pushed to order 0, if $r_1 + r_2 > 2$, and to order 2, if $r_1 + r_2 = 2$.

Looking for instance at the operators that do not contain any field derivatives, we see that they must have the form

$$\lambda_h^{r_1, r_2} \int dx(\psi_x^+)^{r_1}(\psi_x^-)^{r_2} , \qquad (15.4)$$

with

$$\lambda_h^{r_1, r_2} = \sum_{s=1}^{\infty} \binom{s}{r_1}\binom{s}{r_2} \bar{W}_{h,s}\rho^{s-\frac{1}{2}(r_1+r_2)} ,$$

$$\bar{W}_{h,s} = \lim_{|\Lambda| \to \infty} \frac{1}{\Lambda} \int d\vec{x}d\vec{y} W_{h,s}(\vec{x}, \vec{y}) . \qquad (15.5)$$

3. We insert the representation (13.14) in (15.5) and we collect the terms proportional to F_{40}, F_{21}, F_{20}, and F_{02}. Their coefficients are written $\frac{p_0^2 \rho}{2m}$ times $\tilde{\lambda}_h$, $\tilde{\mu}_h$, $-\gamma^{2h}\tilde{\nu}_h$ and $\tilde{Z}_h$, respectively. And (15.5) implies immediately that

$$\frac{p_0^2 \rho}{2m}\tilde{\lambda}_h = \frac{1}{4}\sum_{s=2}^{\infty}\bar{W}_{h,s}\rho^s\sum_{r=0}^{4}(-1)^r\binom{s}{r}\binom{s}{4-r},$$

$$\frac{p_0^2 \rho}{2m}\tilde{\mu}_h = -\frac{1}{\sqrt{2}}\sum_{s=2}^{\infty}s(s-1)\bar{W}_{h,s}\rho^s, \tag{15.6}$$

$$\frac{p_0^2 \rho}{2m}\tilde{Z}_h = \frac{1}{2}\sum_{s=2}^{\infty}s(s-1)\bar{W}_{h,s}\rho^s = -\frac{1}{\sqrt{2}}\mu_h.$$

The full coefficient of $F_{02} = \int(\chi_x^+)^2 dx$ is, therefore, $\tilde{Z}_h + \gamma^{2h}\tilde{\nu}_h$, if $h < 0$.

Furthermore, it is clear that $\tilde{\lambda}_h$, $\tilde{\mu}_h$, $\tilde{\nu}_h$, $\tilde{Z}_h + \gamma^{2h}\tilde{\nu}_h$ are identical to the corresponding constants λ_h, μ_h, ν_h, Z_h defined previously. Hence,

$$Z_0 = -\frac{1}{\sqrt{2}}\mu_0, \qquad Z_h = -\frac{1}{\sqrt{2}}\mu_h + \gamma^{2h}\nu_h \quad \text{if } h < 0, \tag{15.7}$$

is proved.

We are now ready to study the flow of the running couplings and renormalization constants $r_h \equiv (\lambda_h, \nu_h, Z_h, A_h, B_h, E_h)$; the μ_h is eliminated from the scene by (15.7).

As we repeatedly anticipated, we shall study the problem only to the leading nonlinear order. At first it is not really clear what this means. Note that in fact we have two small parameters: $\varepsilon = \lambda\hat{v}(\vec{0})\rho 2mp_0^{-2}$, and λ itself.

We would like to take ε fixed and λ small. But this is not possible without violating the assumptions that made possible approximating (13.2) with the local interaction (13.5), i.e., $\rho p_0^{-3} \ll 1$.[2] This forces us to take ρ fixed, and the only small parameter is λ (or ε equivalently, see (14.13)).

The strategy will be the following. Considering ε as a parameter independent on λ, we show that the first h_0 integrations, with h_0 defined by

$$\gamma^{2h_0} \overset{def}{=} 4\varepsilon, \tag{15.8}$$

---

[2] Dropping such an assumption would force us to study a nontrivial ultraviolet problem, hence there is not much to gain in doing so, if one is interested in studying the simplest case of Bose condensation.

can be performed producing an effective potential $V^{(h_0)}$, with running couplings λ_h, ν_h, μ_h (hence Z_h) still of order ε and

$$|\lambda_h| < 2\lambda_0, \quad |\nu_h| \leq 2\lambda_0, \quad |Z_h| < 2\gamma^{2h}, \tag{15.9}$$

$$|A_h| \leq \frac{1}{2}, \quad |\gamma^{2h} B_h| \leq \frac{1}{2}, \quad |E_h - 1| \leq \frac{1}{2}. \tag{15.10}$$

This is done by a straightforward use of perturbation expansions in the bare constant $\lambda_0 = \varepsilon/4$ (see (14.13)). The results are uniform in ε because the term in ε in the determinant of $G_{(0)}(k)$ (see (13.18)), i.e., $4\varepsilon \frac{p_0^2}{2m} \frac{\vec{k}^2}{2m}$, is small compared to $(\frac{\vec{k}^2}{2m})^2 + k_0^2$, as long as $\frac{\vec{k}^2}{2m} > 4\varepsilon \frac{p_0^2}{2m}$, i.e., as long as we are on scale $h \geq h_0$.

Note that h_0 is the scale below which the Bogoliubov model propagators become symmetric in spacetime, i.e., its dependence on k_0 and $\vec{k}$ is essentially via $k_0^2 + \varepsilon v_0^2 \vec{k}^2$.

In order to understand, assuming (15.10), that the running couplings λ_h, ν_h stay small all the way down to the scale h_0, where the propagators start scaling "correctly," one notes that for $h \geq h_0$ one can easily check that the rescaled propagators, i.e., those for the dimensionless fields $\overline{\chi}^{(h)}$ in (14.8), are bounded by

$$|\tilde{g}_h^{\sigma_1 \sigma_2}(x)| \leq \gamma^{-\frac{1}{2}(\sigma_2 + \sigma_2)h} f(\gamma^h x^0, \vec{x}), \tag{15.11}$$

where f is a rapidly decaying function.

This means that in this first bunch of integrations the power counting is somewhat different, i.e., there are substantial corrections with respect to the asymptotic power counting, which we shall discuss further on.

Consider a graph contributing to the flow equation for λ_h, or μ_h (i.e., Z_h, by (15.7)) (the case of ν_h is analogous and in some sense easier, as ν_h is a relevant operator and we have the freedom of choice of ν_0 to enforce smallness of ν_h), and denote n_σ the number of external lines of type χ^σ, n_λ the number of λ-vertices, and n_μ the number of μ-vertices.

To estimate the contribution of the graph, we see that the bound (15.11) implies that each internal half-line of type χ^σ gives a contribution $\gamma^{-h\sigma/2}$ (on top of the "normal power counting"), while each one of the $(n_\lambda + n_\mu - 1)$ integrations that one has to perform to evaluate the local part of the graph gives a contribution proportional to γ^{-h} (on top of the "normal power counting").

Hence, up to a constant, the graph (which with a normal power counting would, by construction, be a scale independent constant times the appropriate power of the couplings) can be bounded by

$$|\lambda_h|^{n_\lambda}|\mu_h|^{n_\mu}\gamma^{h[\frac{1}{2}(4n_\lambda+2n_\mu-n_-)-\frac{1}{2}(n_\mu-n_+)-(n_\lambda+n_\mu-1)]} =$$
$$= |\lambda_h|^{n_\lambda}|\mu_h|^{n_\mu}\gamma^{h[n_\lambda-\frac{1}{2}n_\mu+\frac{1}{2}(n_++n_-+2)]} \,. \tag{15.12}$$

If we denote N_L the number of independent loops of the graph (equal to the number of propagators $\frac{1}{2}(4n_\lambda+3n_\mu-n_+-n_-)$ minus the "minimum number of propagators" necessary to form a connected graph, i.e., $n_\lambda+n_\mu-1$), $N_L = n_\lambda+\frac{1}{2}n_\mu-\frac{1}{2}(n_++n_-)+1$, and if we use the (15.7), i.e., $Z_h = -\frac{1}{\sqrt{2}}\mu_h$ (neglecting for simplicity the very small $\gamma^{2h}v_h$), we can write the bound (15.12) in the form

$$|\lambda_h|^{n_\lambda}(Z_h^2\gamma^{-2h})^{n_\mu/2}\gamma^{h(N_L+n_+)} \le |\lambda_h|^{n_\lambda}Z_h^{n_\mu/2}\gamma^{h(N_L+n_+)} \,, \tag{15.13}$$

having used (15.9) as well.

The contributions of the irrelevant terms are easily seen to be smaller by at least a power of γ^h (because one should recall that (15.11) holds for the propagators already "normally scaled").

The contributions from the graphs containing n_ν v_h vertices are smaller by a factor ε^{n_ν}, if $|v_h| = O(\varepsilon)$ (which can be imposed, if (15.9) stays true). In fact, each v_h vertex can appear only as an insertion in an internal $\tilde{g}^{--}$ propagator; hence it produces an increase of two in the number of internal half-lines of type χ^- and an increase of one in the number of integrations. But this does not change the power of γ^h in (15.12), so that each v_h vertex changes the bound (15.13) only by a factor $\varepsilon \le \gamma^{2h}$.

Note that the contribution of a graph to the beta function is obtained by calculating its value at zero momentum of the external lines. Therefore no graph with $N_L = 0$ can contribute, because the single scale propagator vanishes at zero momentum. Furthermore, we know that $\lambda_0 = \frac{\varepsilon}{4} = \frac{Z_0}{4}$, so that the variations of Z_h, from (15.13), are bounded by a series with sum of order $O(\lambda_0)$ if $\lambda_h \le 2\lambda_0$ (see Remark below). This is proved, formally, by induction because (15.13) implies that the dominant contributions arise from the one-loop graphs; hence one expects that there is a constant c such that

$$|\lambda_{h-1} - \lambda_h| \le c\lambda_0^2\gamma^h$$
$$|Z_{h-1} - Z_h| = \frac{1}{\sqrt{2}}|\mu_{h-1} - \mu_h| \le c\lambda_0^{3/2}\gamma^{2h} \,, \tag{15.14}$$

which implies $|\lambda_h| \le 2\lambda_0$ and $Z_h \le 2Z_0 = 2\gamma^{2h_0} \le 2\gamma^{2h}$ for $h \ge h_0$, if λ_0 is small enough.

Therefore, if the relations (15.10) hold for $h > h_0$, the relations (15.9) follow.

The discussion still requires the analysis of the relations (15.10). For this purpose note that all the graphs contributing to B_{h-1} and A_{h-1} are graphs with $n_- = 2, n_+ = 0$: their contribution has the form $\int dx x_0^2 W_G(x)$ and $\int dx \bar{x}^2 W_G(x)$, respectively. Hence, by (15.13) (and (15.11)) we see that the bound on the variation $A_{h-1} - A_h$ is $O(\lambda_0 \gamma^h)$, while that of $B_{h-1} - B_h$ has to be multiplied by the extra factor γ^{-2h} (because $(x^0)^2$ is scaled in (15.11) by γ^{2h}). Hence,

$$|B_{h-1} - B_h| \leq c\lambda_0 \gamma^{-h}, \qquad |A_{h-1} - A_h| \leq c\lambda_0 \gamma^h . \qquad (15.15)$$

In a similar way one can prove that

$$|E_{h-1} - E_h| = 2|d_h| \leq c\lambda_0 \gamma^h . \qquad (15.16)$$

The relations (15.10) easily follow, if λ_0 is small enough.

Remark. To repeat once more: the above is just a consistency check on the expansions in powers of λ_h, ν_h. In order to make it into a rigorous proof one cannot just use the perturbative series as we expect them *not to be convergent* but only asymptotic. As in chapter 8 one can only hope to prove an $n!$ bound on the sum of all graphs with n vertices.

All that the above preliminary discussion shows is that (to no one's surprise) it is sufficient to study the running couplings flow in the region $h < h_0$. The latter is the region where, if, for some constants c, c_1

$$\frac{1}{2} \leq 1 + 4A_h \leq c, \qquad cZ_h \leq E_h^2 + 16 B_h Z_h \leq c_1 Z_h ,$$
$$0 < Z_h < 2Z_0 \equiv 2\varepsilon , \qquad (15.17)$$

the rescaled propagators $\tilde{g}_h^{--}(x)$ and $\tilde{g}_h^{-+}(x)$ are essentially independent of h, i.e., they can be bounded by a rapidly decaying function $f(x)$, uniformly in A_h, B_h, Z_h, E_h, verifying (15.17) (see appendix 6). On the contrary, the propagator $\tilde{g}_h^{++}(x)$ can be bounded by $f(x)/Z_h$.

The previous properties of the rescaled propagators and the identity (15.1) imply that, in the bound of a generic graph contributing to the beta function, two μ_h vertices are essentially equivalent to one λ_h vertex. A further simple analysis allows us to prove that, if we want to keep in each flow equation only the leading terms, then we have to consider only one loop graphs without ν_h vertices.

There are very "few" graphs with one loop; therefore the calculation is straightforward, but quite long. Here we report the result, while the details are exposed in appendix 6. Neglecting terms of order γ^{2h} (which essentially come from the corrections to the scaling of the propagators, which become quickly scale independent), and the contributions from the graphs with more than one loop, one finds after using (15.7) to eliminate μ_h from the row result (see appendix 6) and after setting $\zeta_h \equiv 4Z_h$,[3]

$$\lambda_{h-1} = \lambda_h - 36\beta_{2,h}\zeta_h^2\left(\lambda_h - \frac{\zeta_h}{24}\right)^2 ,$$

$$\zeta_{h-1} = \zeta_h - \frac{1}{4}\beta_{2,h}\zeta_h^4 ,$$

$$E_{h-1} = E_h - \frac{1}{4}\beta_{2,h}\zeta_h^3 E_h , \qquad (15.18)$$

$$A_{h-1} = A_h ,$$

$$B_{h-1} = B_h + \frac{1}{16}\beta_{2,h}\zeta_h^2 E_h^2 ,$$

$$\nu_{h-1} = \gamma^2 \left(\nu_h + \beta_{1,h}(6\zeta_h\lambda_h - \frac{1}{4}\zeta_h^2)\right) ,$$

where

$$\beta_{2,h} = \frac{\log\gamma}{8\pi^2\sqrt{a_h^3 b_h}}\frac{p_0^3}{\rho}, \qquad \beta_{1,h} = \frac{1-\gamma^{-2}}{8\pi^2(\sqrt{a_h b_h}+a_h)}\frac{p_0^3}{\rho} , \qquad (15.19)$$

$$a_h = \zeta_h(1+4A_h), \qquad b_h = (E_h^2 + 4\zeta_h B_h) ,$$

and the terms in $\gamma^{2h}\nu_h$ coming from the application of (15.7) to eliminate μ_h have been dropped, as we plan to impose that $\nu_h \xrightarrow[h\to-\infty]{} 0$, so that they are at least of order $O(\gamma^{2h})$.

It is very easy to analyze this flow, under the conditions (15.17), implying that $\beta_{2,h}$ is of order ζ_h^{-2}. In fact, this observation is sufficient to prove that $\zeta_h = O(1/|h|)$ for $h \to -\infty$. But this property has to be true also for E_h, since, by (15.18),

$$\frac{E_{h-1} - E_h}{E_h} = \frac{\zeta_{h-1} - \zeta_h}{\zeta_h} . \qquad (15.20)$$

It is now very easy to check that

$$B_h \xrightarrow[h\to-\infty]{} B_{-\infty} > 0 , \qquad (15.21)$$

---

[3] The notation should not mislead the reader into confusing the present ζ_h with the renormalization constant in (14.12).

while A_h stays constant (indeed a small constant, roughly equal to its value on scale h_0).

Finally, if we define

$$c_0 = \frac{p_0^3 \log \gamma}{16\pi^2 \rho \sqrt{B_{-\infty}(1+4A_{-\infty})^3}} , \qquad (15.22)$$

we see that the first two equations in (15.18) can be written, for $h \to -\infty$, in the form

$$\lambda_{h-1} = \lambda_h - 36c_0\zeta_h^2 \left(\frac{\lambda_h}{\zeta_h} - \frac{1}{24}\right)^2 ,$$
$$\zeta_{h-1} = \zeta_h - \frac{c_0}{4}\zeta_h^2 \qquad\qquad\qquad (15.23)$$

(where the fact that the first equation sums to a perfect square should come from some a priorireason that we have not been able to discover).

The discussion of the above equations is elementary and, starting from initial data $\zeta_0 = 4\varepsilon$, $\lambda_0 = \zeta_0/16$ (or any others close to them), the result is that, if ν_0 is chosen so that ν_h is bounded uniformly in h, then, asymptotically,

$$Z_h = \bar{c}|h|^{-1}, \quad \lambda_h = \frac{1}{4}Z_h, \quad \nu_h = O(\lambda_h) , \qquad (15.24)$$

if $\bar{c}$ is a suitable constant (ε independent).

Note that these results are consistent with (15.17), which can then be proved inductively, together with (15.24).

At this point it is very easy to check that all the neglected terms in the beta function are at least of order $1/|h|^2$. Hence they cannot change in a substantial way the asymptotic properties of the flow (up to convergence problems; see the remark above); only the values of $A_{-\infty}$, $B_{-\infty}$, and c depend on them, and $A_{-\infty}$ has to be a small number (of order ε). Note that this last observation is important, in order to be sure that the first condition in (15.17) is preserved, since we do not have a control on the sign of $A_{-\infty}$.

The main consequence of the previous discussion is that, for $k \to 0$ (that is for $h \to -\infty$), the model is Gaussian (asymptotic freedom) and the pair Schwinger function of the fields $\psi^\pm$ behaves as

$$\tilde{S}_{-+}(k) = -\tilde{S}_{--}(k) = -\tilde{S}_{++}(k) \simeq \frac{p_0^2}{16m\rho B_{-\infty}} \frac{1}{k_0^2 + c_{-\infty}^2 \vec{k}^2} , \qquad (15.25)$$

where the sound speed $c_{-\infty}$ is given by

$$c^2_{-\infty} = \frac{(1+4A_{-\infty})v_0^2}{16B_{-\infty}} = v_B^2[1 + O(\sqrt{\varepsilon})] , \qquad (15.26)$$

where $v_B^2 = \varepsilon v_0^2$ is the square sound speed in the Bogoliubov model. In fact, the bound (15.15) implies that A_{h_0} is of order ε, and B_{h_0} is of order $\sqrt{\varepsilon}$, (15.16) implies that $1 - E_{h_0}$ is of order ε and (15.14) implies that $Z_{h_0} = \varepsilon[1 + O(\sqrt{\varepsilon})]$; moreover, by (15.20), for $h < h_0$,

$$Z_h \simeq \frac{Z_{h_0}}{E_{h_0}} E_h = \varepsilon E_h[1 + O(\sqrt{\varepsilon})] , \qquad (15.27)$$

and, by (15.18),

$$\frac{B_{h-1} - B_h}{E_{h-1} - E_h} = -\frac{E_h}{16Z_h} = -\frac{1}{16\varepsilon}[1 + O(\sqrt{\varepsilon})] , \qquad (15.28)$$

implying that $B_{-\infty}$ is of order $1/\varepsilon$, since $E_{-\infty} = 0$, so that

$$B_{-\infty} = \frac{1}{16\varepsilon}[1 + O(\sqrt{\varepsilon})] . \qquad (15.29)$$

A Brief Historical Note

At the request of the series editor, Prof. Arthur Wightman, we provide here a very personal attempt at some historical considerations on the development of the renormalization group ideas. We do not feel particularly qualified to do history seriously because our knowledge of the literature is very partial. Therefore, what follows really represents the history of our cultural evolution through the subject. The papers and works that we quote are the ones that we met in the course of our active research, either in trying to solve some of the problems in which we have been involved or in attending related conferences and schools. Among the latter we mention the Cargese summer school 1973, the Roma conference on mathematical physics of spring 1977, and the Les Houches summer school 1984. And since our background is in mathematical physics it is inevitable that we give our sketch a mathematical physics perspective.

Many years ago (1948), the discovery of renormalizability of QED gave rise to the first form of the renormalization group, as a (semi)group of covariance of the Schwinger functions of a field theory with respect to the renormalization procedures used to construct them (i.e., their formal perturbation expansion).

This approach led to an understanding of how the perturbation theory could be improved by resumming some classes of contributions to the value of a Schwinger function, associated with families of Feynman graphs (e.g., one-loop chains).

The first to realize, to our knowledge, that the renormalization group ideas could be applied to statistical mechanics and in particular to the analysis of critical behavior were Di Castro and Jona-Lasinio [DJ]; their paper, however, did not attempt any concrete calculation (because they "contented" themselves with indicating that, under suitable assumptions, the scaling laws of the critical point of a ferromagnet would be a consequence of renormalizability of φ^4 scalar field theories). The work went largely unnoticed (see [W5]). Sometimes the paper was even ironically commented upon.

At about the same time, an independent and fresh approach to the problem was developed by Wilson. His work on the hierarchical model (the so-called Wilson recursion relation), with its simplicity and clarity, opened up new horizons to many, and led to the success of the ε-expansion devised by Wilson and Fisher and intensively applied to many problems, particularly by the French schools ([IZ]). Many others were convinced that this was the right approach to the critical point theory and to the renormalization theory in QFT.

The novel notion of asymptotic freedom was the really new information beyond the work of Dyson, Feynman, and Schwinger; its spectacular application to QFT in the frame of noncommutative gauge theories was one of its major successes ([Ho]).

Some predictions were quite striking, for instance the prediction of the values of the critical exponents in $d \geq 4$ ferromagnets, though they were met with skepticism in the mathematical physics community.

By 1973 the subject began to attract the attention of mathematical physicists with the work of [BS1] on Dyson's hierarchical model (a model for which the Wilson recursion relations are exact). A few years later came the rederivation of the ultraviolet stability in scalar QFT by the Roma school, which dealt for the first time with the problem of treating the remainders of perturbation theory (the large field problem) in QFT by literally computing the effective potentials that were so widely used in the formal treatments of QFT and controlling their sizes. In the hierarchical case the ideas were essentially the same as the ones in [BS1], [BS2]. They were carried over to the nonhierarchical QFT (in $d = 3$ space time dimensions) ([Rm2]).

Not only did this establish a connection with all the works on QFT that were done until that time in constructive field theory by Nelson, Glimm-Jaffe, and Guerra ([N], [GJ], [Gu]), but it permitted us to realize that the ideas and methods in the two fields were deeply similar. The Wilson approach to QFT was thus quite clearly related, on a technical level, to the corpus of results on QFT emanating from the classical formulation of the Garding-Wightman axioms ([WG]), which had attained their highest levels only in the previous years, in parallel to the development of the Wilson theory, with the works of Glimm, Jaffe and Spencer, Feldman and Osterwalder, and Magnen and Seneor ([GJS], [FO], [MS]) and that would still produce important results (De Calan and Rivasseau, [DCR]).

The work on Anderson's localization provided independent evidence that the new ideas were suited to the analysis of many other problems.

The number of results obtained in mathematical physics by "renormalization group methods" quickly became very large. The summer school of Les Houches gives a good cross section of the status at the time. Besides the new renormalization theory approaches that stemmed out of Hepp's work ([H]) in [GN1], [GN2], [G1], there were also the important results of Gawedski and Kupiainen on φ_4^4 infrared (critical point of ferromagnets) ([GK3]).

Other branches of physics profited as well, as in the theory of breakdown of invariant tori in classical mechanics by McKay and by Shenker and Kadanoff ([McK], [SK]).

The fundamental work of Russo ([Ru]) on the $d = 2$ Ising model and its subsequent (independent) developments by Aizenman [Ai1] and Higuchi [Hi] gave rise to developments that led Aizenman [Ai2] to confirm the renormalization group predictions about the critical point behavior of ferromagnets in $d \geq 4$. The triviality conjecture about φ_4^4 also became much better understood (Fröhlich [Fr] and Aizenman [Ai2]), although today it is essentially still as open (and challenging) as it was left after the just-mentioned works.

The present status of the field is not very satisfactory: there are quite a few papers whose claims have not been checked by anyone but the authors themselves. Some of them are quite important.

It is unclear why we are in such a situation. Basically this might be because the papers have become very technical, and, except in a few cases, the results are "of little physical interest" because they are allegedly well understood by physicists. Therefore no one appears to be interested in devoting the huge amount of time necessary to the checking of the results.

This means that often the same results are repeated, with the authors apparently not even noticing that what they are doing was already done elsewhere. This situation is not all bad, as the more a result is independently derived, the more reliable it becomes and the less likely it is to be forgotten.

This also means that another mechanism might be operating: people do not like results that cannot be tested quickly for correctness. Neither do they like results that can be obtained quickly from results they did not test themselves. This leads necessarily to the attempt on the part of the authors to write self-contained papers that become longer and longer and, therefore, more and more unreliable.

There seems to be no way out of this in the foreseeable future; we hope that the new generations will consider the renormalization group approaches as natural as we find the Feynman graphs in perturbation expansions today, so that papers can finally be written without going back to the origins each time.

But it is necessary also that scientists give up the aristocratic view that good science is "simple": there are matters that might be intrinsically involved and difficult and require long analysis. We may have just exhausted the "simple" problems in classical and quantum physics, at least the ones that our generation is willing to consider simple. And we should adapt to the new knowledge that has been created.

Of course the authors bear the responsibility of making their work accessible, and they should make all possible efforts to try to simplify their work, thus helping both themselves and others in the necessary checks.

Bibliographical Notes

Chapter 1

Other problems treated with the renormalization group are the theory of the Kosterlitz-Thouless transition by Fröhlich and Spencer ([FS]), the theory of the dipole gas ([GK4]), and the Kondo effect ([W6]). A general unified exposition of the methods and of the physical ideas is in [P].

Chapter 3

The extensions of the Fermi liquid ground state to dimensions higher than 1 are formally quite easy. The extension of the notion of field and the related functional integral is discussed in [BG1], [BG2]. An alternative approach is in [FT1], [FT2]. However, at this moment (1994) the results still seem to be quite incomplete. The definition of "complete result" that we use is a result that can be formulated without referring to technical problems generated by the formalism, hence, that which can be explained by referring only to the system under study (i.e., as a property of the Hamiltonian). Recently, a number of interesting papers and ideas have appeared that allow us to formulate a rather precise set of conjectures ([Po2], [Sh], [We], [FMRT1], [FMRT2]) for Fermi liquids; a related paper is [BG2]. Such results, which cannot yet be considered as mathematically rigorous, develop a picture of the superconductivity phenomenon in Fermi liquids in dimension $d = 2, 3$ by relating it to the properties of an N component system of interacting fermions in the limit $N \rightarrow \infty$. The lack of mathematical completeness is not the reason we do not treat such cases here. In fact, we have already explained that this book is not meant at all to be a mathematical treatise. In this book we prefer to discuss another problem at a comparable level of development, namely the theory of the $d = 3$ Bose gas, because it is relatively newer and more controversial, or at least as controversial (and we control it better from a technical point of view). On the other extreme are recent works on the one-dimensional (spinning) Fermi liquids on a periodic potential in $d = 1$, which we do not discuss for lack of space and because they were completed after the present book was essentially laid down (long ago, in fact), but which complete considerably the picture that stems out of the treatment of the spinless fermions in $d = 1$ presented here ([BM]).

The book by Rivasseau ([R]) is a presentation of similar problems with more attention given to the mathematical aspects. It also develops in detail the point of view on the renormalization group developed by the French school.

Chapter 6

The key to the development of the notions of relevant, marginal, and irrelevant operators are the works of Wilson on the hierarchical model. They were then developed to mathematical theories on Dyson's hierarchical model by Bleher and Sinai ([BS1] [BS2]), and summarized and extended in the book by Collet and Eckmann ([CE]). The application of Wilson's ideas to quantum field theory appeared shortly afterwards and in [G2], [Rm1], and [Rm2], where the basic ideas on how to control the "large fields" problem (i.e., on how to obtain remainder estimates on the perturbation theory expansions) are developed.

Chapter 7

The notion of asymptotic freedom, due to Wilson, is perhaps the key notion that was missing in the original renormalization theory of Dyson, Feynman, and Schwinger. It allows us to put a finer distinction between theories that are renormalizable.

Chapter 8

The beta function essentially appears in the work by Wilson and Fisher ([WF]). It was developed to become a very widely used tool by many (see [IZ] for an early summary). The formalism developed here is the so-called *tree formalism*: it goes back to Hepp ([H]). The interpretation in terms of beta function presented here is due to Gallavotti and Nicolò ([GN1], [GN2]). Rather different uses of the same ideas have been made in the theory of the critical point in ferromagnets by Aizenman ([Ai2]). The triviality conjecture of φ_4^4 discussed by Fröhlich ([Fr]) and Aizenman ([Ai2]) also rests on similar ideas or initiates related ideas.

Chapter 10

The anomalous dimension arose in the work of Wilson and Fisher ([WF]). The work of Felder ([Fe]) and then Gawedski and Kupiainen ([GK2]) began to introduce the theme in the mathematical physics lit-

erature. In the latter papers, however, the anomalous dimension is prescribed a priori, not determined, as in [WF], "dynamically." The discussion here is only about the dynamical anomalous dimension; it follows
a scheme that was explained to one of us (GG) by Felder in the case of
a scale decomposition with sharp cutoffs separating the various scales.
Felder's point of view is not published; the present exposition is quite
faithful if performed in the sharp cutoff case (such a cutoff is in fact very
convenient for calculations once one understands that the discontinuities
in the cutoff functions do not produce divergencies in the calculation, as
discussed for instance in chapter 15; see also [P].

The dynamical anomalous dimension appears, with an attempt at
deriving it on mathematically rigorous grounds (in a case in which it is
actually present), in the thesis of Da Veiga ([DaV]) and in [BGPS]. The
treatment in the latter paper is at the basis of the exposition presented
here.

Chapter 11

The Luttinger model was, in fact, solved exactly by Lieb and Mattis
([ML]). The theory of the spinless $d = 1$ Fermi liquid was basically
explained by Tomonaga ([T]). The use of the properties of an exactly
soluble model to infer properties of the beta function for a nonsoluble
one is an idea developed in [BG1], [BGM], and brought to completion
in [BGPS]. It has been proposed to be useful in the theory of $d = 2$
Fermi liquids by Anderson ([A1] [A2]; see also [G3]). The idea has
not yet developed into a mathematical theory. The first theory of the
$d = 1$ Fermi liquid, including spinning fermions, is due to Sólyom ([So]);
the works of [BGPS], [BM] are, essentially, a mathematically rigorous
version of it.

Chapter 12

This chapter represents our version of the ε expansion of [WF] and
[WK]. It is based on the version of the anomalous dimension theory in
chapter 10. A more classical version can be found in [P].

Chapter 13

The theory of the Bose gas is in a quite primitive stage. There do not
seem to be treatments based on the renormalization group — at least no
noncontroversial treatments. Here we propose one more due to one of us

([B]). The standard ideas can be found in ([ADG]). We do not discuss the $d = 1$ exactly soluble cases, treated by Lieb and Liniger ([LL]), who initiated a wide research on the theme, culminating in the papers by Vaidya and Tracy ([VT]), and by Jimbo, Miwa, Mori, Sato ([JMMS]). There are few mathematically rigorous results: among them the papers by Ginibre ([Gi]) and by Kennedy, Lieb and Shastri ([KLS]) are certainly prominent. The theory of Bogoliubov ([Bo]) is well accepted as a first approximation. It seems unlikely that it is basically wrong: our logarithmic corrections to it are, in some sense, minor variations.

Appendix 1
The Free Fermion Propagator

In the free case the pair correlation function (see (3.2)) can be written

$$S(\vec{x} - \vec{y}, t - t') = \begin{cases} \langle \psi^-_{\vec{x},t} \psi^+_{\vec{y},t'} \rangle & t > t' \\ -\langle \psi^+_{\vec{y},t'} \psi^-_{\vec{x},t} \rangle & t \le t' , \end{cases} \tag{A1.1}$$

where $|t - t'| < \beta$, $\langle A \rangle \equiv \mathrm{Tr}\,(Ae^{-\beta H})/Tr(e^{-\beta H})$, H is the second quantization version of the operator (3.1) with $\lambda = 0$:

$$\psi^\pm_{\vec{x},t} \equiv e^{tH} \psi^\pm_{\vec{x}} e^{-tH} = \frac{1}{\sqrt{L^d}} \sum_{\vec{k}} e^{\pm i\vec{k}\cdot\vec{x}} e^{\pm \varepsilon(\vec{k})t} a^\pm_{\vec{k}} \equiv$$

$$\equiv \frac{1}{\sqrt{L^d}} \sum_{\vec{k}} e^{\pm i\vec{k}\cdot\vec{x}} a^\pm_{\vec{k},t} \tag{A1.2}$$

and

$$\varepsilon(\vec{k}) = \frac{\vec{k}^2 - p_F^2}{2m} . \tag{A1.3}$$

One notes that $\langle a^\pm_{\vec{k},t} a^\mp_{\vec{k}',t'} \rangle = 0$ unless $\vec{k} = \vec{k}'$. Supposing $\beta > t > t' > 0$, it is

$$\langle a^\pm_{\vec{k},t} a^\mp_{\vec{k}',t'} \rangle = \frac{\sum_E \langle E| e^{-(\beta-t)H} a^\mp_{\vec{k}} e^{-(t-t')H} a^\pm_{\vec{k}} e^{-t'H} |E\rangle}{\sum_E e^{-\beta E}} , \tag{A1.4}$$

where $|E\rangle$ denotes a generic eigenvector $|E\rangle = |\vec{k}_1, \ldots, \vec{k}_n\rangle$ of the operator H consisting of n occupied momentum levels $\vec{k}_1, \ldots, \vec{k}_n$. Then one immediately finds

$$\langle a^-_{\vec{k},t} a^+_{\vec{k},t'} \rangle = \frac{e^{-(t-t')\varepsilon(\vec{k})}}{1 + e^{-\beta\varepsilon(\vec{k})}}, \qquad t - t' > 0 . \tag{A1.5}$$

Likewise, if $t < t'$

$$\langle a^+_{\vec{k},t} a^-_{\vec{k},t'} \rangle = \frac{e^{-(\beta+(t-t'))\varepsilon(\vec{k})}}{1 + e^{-\beta\varepsilon(\vec{k})}}, \qquad \beta + (t - t') > 0 . \tag{A1.6}$$

Hence $S(\vec{\xi}, \tau)$ in (A1.1) is given by

$$\frac{1}{L^d} \sum_{\vec{k}} e^{-i\vec{k}\cdot\vec{\xi}} \left(\chi(\tau > 0) \frac{e^{-\tau\varepsilon(\vec{k})}}{1 + e^{-\beta\varepsilon(\vec{k})}} - \chi(\tau \le 0) \frac{e^{-(\beta+\tau)\varepsilon(\vec{k})}}{1 + e^{-\beta\varepsilon(\vec{k})}} \right), \quad (A1.7)$$

where the sum runs over the $\vec{k}$'s such that $e^{-i\vec{k}L} = 1$, $\chi(\cdot)$ denotes the characteristic function of the event described in the χ argument and $|\tau| < \beta$.

The expression in parentheses can be written as

$$\frac{1}{\beta} \sum_{k_0} \frac{e^{-ik_0\tau}}{-ik_0 + \varepsilon} \equiv -\frac{1}{\beta} \sum_{k_0} \frac{e^{-ik_0(\beta+\tau)}}{-ik_0 + \varepsilon}, \quad (A1.8)$$

where the sum is over the k_0's such that $e^{-ik_0\beta} = -1$.

In fact, if $\tau > 0$, the first expression of (A1.8) can be written as

$$\oint \frac{dz}{2\pi} \frac{e^{-iz\tau}}{(-iz + \varepsilon)(1 + e^{-i\beta z})}, \quad (A1.9)$$

where the contour runs just below the real axis from $-\infty$ to $+\infty$ and comes back just above it from $+\infty$ to $-\infty$, picking up the residues at the poles where $e^{-i\beta z} = -1$. On the other hand, if $\beta > \tau > 0$, the above two contours can be made to recede to ∞ and the integral is minus the residue at $z = -i\varepsilon$ (because the contour below the real axis is deformed to a clockwise circle around $-i\varepsilon$); hence the value of (A1.9) is $e^{-\varepsilon\tau}(1 + e^{-\beta\varepsilon})^{-1}$.

If $\tau \le 0$ we repeat the same argument by using the second expression in (A1.8), and that $\beta > \beta + \tau > 0$. We find that for $\tau \le 0$ the value of (A1.8) is $-e^{-\varepsilon(\beta+\tau)}(1 + e^{-\beta\varepsilon})^{-1}$.

Finally we can compute the density,

$$\rho = -S(\vec{x}, 0^-) = \langle \psi_{\vec{x},0}^+ \psi_{0,0}^- \rangle = \frac{1}{L^d} \sum_{\vec{k}} e^{-i\vec{k}\cdot\vec{x}} \frac{e^{-\beta\varepsilon(\vec{k})}}{1 + e^{-\beta\varepsilon(\vec{k})}}, \quad (A1.10)$$

and we see that, if $L \to \infty$, $\beta \to \infty$, it is

$$\rho = \frac{1}{(2\pi)^d} \int d^d\vec{k} \, e^{-i\vec{k}\cdot\vec{x}} \chi(\varepsilon(\vec{k}) \le 0), \quad (A1.11)$$

which defines the Fermi sphere in the free case.

Appendix 2
Grassmannian Integration

Consider the algebra generated by the identity and by a family of *pairwise anticommuting* operators ε_k^σ, A_k^σ, where the labels σ are ± 1 and the k's are $k = (k_0, \vec{k}) \in \mathbf{R}^{d+1}$, verifying

$$e^{-ik_0\beta} = -1, \qquad e^{-i\vec{k}L} = +1 , \qquad (A2.1)$$

and pairwise anticommuting is required also for the pairs consisting of identical operators. The ε, A operators will be called *Grassmannian variables.*

It is most convenient to think of the A, ε as concrete objects by using a representation on a Hilbert space h. The best Hilbert space is probably a countable tensor product of two-dimensional spaces $\mathbf{C}^2$: $h = \otimes_{j=1}^\infty \mathbf{C}^2$ based on the vector $|\Omega\rangle$ ("vacuum") which is an infinite tensor product of vectors $\begin{pmatrix} 0 \\ 1 \end{pmatrix}$. Then we order (absolutely arbitrarily) the variable labels, by replacing each of them with an integer label $j = 1, 2, \ldots$ and set the j–th, Grassmannian variable, to be

$$[\prod_{i<j} \sigma_i^z]\sigma_j^+ , \qquad (A2.2)$$

where σ^z, σ^+ are the usual Pauli matrices.

Hence the Grassmannian variables can be regarded as a set of creation operators (just creation and no annihilation) on a Fock space (the vacuum being the vector $|\Omega\rangle$ used to define the tensor product).

The A, ε variables are norm 1 operators on h. They will be used to define the *euclidean field* with *ultraviolet cutoff* on scale γ^{-U} and *infrared cutoff* on scale γ^{-R}, ψ_x^σ, and the *external field* φ_x^σ as

$$\psi_x^\sigma = \sum_k \frac{e^{i\sigma kx}}{\sqrt{\beta L}} \frac{(e^{-k^2\gamma^{-2U}} - e^{-k^2\gamma^{-2R}})^{1/2}}{\sqrt{-ik_0 + e(\vec{k})}} A_k^\sigma ,$$

$$\varphi_x^\sigma = \sum_k \frac{e^{i\sigma kx}}{\sqrt{\beta L}} \varepsilon_k^\sigma , \qquad (A2.3)$$

where $x = (\vec{x}, t)$, $k^2 = k_0^2 + e(\vec{k})^2$ and $U, -R$ are large positive numbers, while γ is an arbitrary scale parameter (which we take to be $\gamma = 2$, to fix the ideas).

The above operators will be generically denoted $\Phi_x^{\pm}$ and sometimes the fields with cutoffs at U and R will be denoted $\psi_x^{[U,R]\sigma}$, when the cutoff dependence has to be made explicit. They are bounded operators on **h**, because the A's have norm 1.

We call $\mathcal{G}_0$ the algebra generated by the A, ε operators and $\mathcal{G}$ its norm closure in the norm of the Hilbert space **h**. The cutoff fields, of course, are elements of $\mathcal{G}$. It is also convenient to define the norm closed algebra $\overline{\mathcal{G}}$ generated by the operators in $\mathcal{G}$ *together with their adjoints*. In mathematical terms, one sees that $\overline{\mathcal{G}}$ is a C^* algebra.

Consider the elements O of $\mathcal{G}$ that can be written as

$$O = \sum_n \int O_n(x_1, \ldots, y_n) \mathcal{D}_{x_1} \Phi_{x_1}^+ \ldots \mathcal{D}_{y_n} \Phi_{y_n}^- \, dx_1 \ldots dy_n , \qquad (A2.4)$$

where the $O_n(\ldots)$ are the "kernels of O," *antisymmetric in the permutations of the x's or y's*, and $\Phi^{\pm}$ are Grassmannian field operators, and $\mathcal{D}_{x_1} \ldots \mathcal{D}_{x_n}$ are differentiation operations of order bounded by some n_0, for all n. Furthermore, the O_n should be measures (i.e., δ functions are allowed) and

$$|O(\Phi)|_z \equiv \sum_n z^n \int |O_n(x_1, \ldots, y_n)| dx_1 \ldots dy_n < \infty \qquad \forall z > 0 . \ (A2.5)$$

Using that $\overline{\mathcal{G}}$ is a C^* algebra generated by creation and annihilation operators, it is not difficult to see that an element of $\mathcal{G}$ admits at most one representation like (A2.4), (A2.5). Note also that the operators A, ε themselves admit a representation like (A2.4), (A2.5).

Definition: *Let $O(\psi, \varphi)$ be an operator in $\mathcal{G}$ admitting the representation (A2.4), (A2.5). Then O will be called integrable. Let O be written, with $dx \equiv dx_1 \ldots dx_n$, $dy \equiv dy_1 \ldots dy_n$, etc.:*

$$O(\psi, \varphi) = \sum_{m^{\pm}, n^{\pm}} \int dx \, dy \, du \, dv \cdot$$

$$\cdot O(x_1, \ldots, x_{n^-}, y_1, \ldots, y_{n^+}, u_1, \ldots, u_{m^-}, v_1, \ldots, v_{m^+}) \cdot \qquad (A2.6)$$

$$\cdot \left(\prod_{i=1}^{n^-} \mathcal{D}_{x_i} \psi_{x_i}^- \right) \left(\prod_{i=1}^{n^+} \mathcal{D}_{y_i} \psi_{y_i}^+ \right) \left(\prod_{i=1}^{m^-} \mathcal{D}_{u_i} \varphi_{u_i}^- \right) \left(\prod_{i=1}^{m^+} \mathcal{D}_{v_i} \varphi_{v_i}^+ \right) .$$

The integral of $O(\psi, \varphi)$ with respect to ψ will be defined as

$$\Omega(\varphi) = \sum_{m^\pm, n^- = n^+} \int dx\, dy\, du\, dv \cdot$$

$$\cdot O(x_1, \ldots, x_n, y_1, \ldots, y_n, u_1, \ldots, u_{m^-}, v_1, \ldots, v_{m^+}) \cdot \qquad (A2.7)$$

$$\cdot \left[\left(\prod_{i=1}^{n} \mathcal{D}_{x_i} \mathcal{D}_{y_i}\right) \det g^{[R,U]}(x_i - y_j)\right] \left(\prod_{i=1}^{m^-} \mathcal{D}_{u_i} \varphi_{u_i}^-\right)\left(\prod_{i=1}^{m^+} \mathcal{D}_{v_i} \varphi_{v_i}^+\right),$$

where n is the common value of n^+ and n^-, and $g^{[R,U]}(x - y)$ is the propagator

$$g^{[R,U]}(x - y) =$$

$$\frac{1}{\beta L} \sum_k e^{-ik(x-y)} \frac{e^{-2^{-2U}(k_0^2 + e(\vec{k})^2)} - e^{-2^{-2R}(k_0^2 + e(\vec{k})^2)}}{-ik_0 + e(\vec{k})} . \qquad (A2.8)$$

Remarks

1. This is well defined because the representation (A2.6) is unique, and the convergence is assured by the assumption (A2.5) and by the Hadamard inequality (see [GK1]):

$$\left|\mathcal{D}_1 \ldots \mathcal{D}_{2n} \det\left[g^{[R,U]}(x_i, y_j)\right]\right| \leq B_{R,U}^n . \qquad (A2.9)$$

2. The definition above is consistent with an alternative definition that defines the Grassmannian integral by linearity from the following propagator

$$\begin{cases} \int P(d\psi) A_k^- A_{k'}^+ = \delta_{k,k'} , \\[2mm] \int P(d\psi) A_k^+ A_{k'}^- = 0 , \\[2mm] \int P(d\psi) A_k^- A_{k'}^- = 0 , \end{cases} \qquad (A2.10)$$

while the integrals of monomials of higher order are given by the "Wick rule" with propagator (A2.10). This means that the integral of an arbitrary monomial in the A^+ and A^- is obtained by considering the pairings of the monomial factors into pairs with nonzero propagator, and then summing the product of the propagators corresponding to the pairs times a sign $\pm$ equal to the parity of the permutation necessary to bring the elements of the considered pairs next to each other. The φ

fields (and the constants) act as constants in the definition of integral. In fact, it is not difficult to see that the determinants in (A2.9) arise precisely from the summations of the pairings.

3. An obvious extension is the *multiple Grassmannian integrals* of independent Grassmannian variables: suppose that instead of just one we have many A variables, labeled by an extra label α. Calling $\psi^{[\alpha]\sigma}$ the corresponding fields with cutoffs $[U_\alpha, R_\alpha]$, we can consider operators $O(\psi^{[\alpha_1]}, \ldots, \psi^{[\alpha_q]}, \varphi)$ and define the partial integrations as in remark (2) with the propagator

$$
\begin{cases}
\displaystyle \int P(d\psi) A_k^{[\alpha]-} A_{k'}^{[\alpha']+} = \delta_{\alpha\alpha'} \delta_{k,k'} \ , \\[2mm]
\displaystyle \int P(d\psi) A_k^{[\alpha]+} A_{k'}^{[\alpha']+} = 0 \ , \\[2mm]
\displaystyle \int P(d\psi) A_k^{[\alpha]-} A_{k'}^{[\alpha']-} = 0 \ .
\end{cases}
\tag{A2.11}
$$

The integration can be denoted $\int P(d\psi^{[1]}) \ldots P(d\psi^{[2]})$.

And, with reference to remark (3) one can prove the following theorem.

Theorem: Let $\psi^{[1]}, \psi^{[2]}, \ldots, \psi^{[q]}$ be a family of independent Grassmannian fields. Then the integrals of operators depending only on the sum $\psi = \psi^{[1]} + \psi^{[2]} + \ldots + \psi^{[q]}$ verify

$$
\begin{aligned}
\int \prod_{i=1}^{q} P(d\psi^{[i]}) \, O(\sum_i \psi^{[i]}) &= \\
= \int P(d\psi^{[1]}) \left(\int \prod_{i=2}^{q} P(d\psi^{[i]}) \, O(\sum_i \psi^{[i]}) \right) &= \\
= \int P(d\psi) \, O(\psi) \ ,
\end{aligned}
\tag{A2.12}
$$

where ψ is a Grassmannian variable with propagator

$$
g(x - y) = \sum_{i=1}^{q} g^{[i]}(x) \ .
\tag{A2.13}
$$

The above (A2.12) statement can be called *Fubini's* theorem and the (A2.13) makes the above Grassmannian integrations very close to the gaussian integrations of functionals of random fields. The important difference is that *no positive definiteness* is required on the propagators

$g^{[i]}$. The latter is not surprising because the above Hadamard inequality (A2.9) solves all the convergence problems that one has to worry about in the theory of gaussian integrals.

An example of the above analysis is statement (3.6). One checks that the operator $V = \lambda \int_\Lambda \psi_x^+ \psi_x^- \psi_y^+ \ \psi_y^- \ \delta(x_0 - y_0) \ v(\vec{x} - \vec{y}) dx dy$ has the form (A2.4), (A2.5) and that $e^{-V(\psi)} \psi_{x_1}^{\sigma_1} \ldots \psi_{x_n}^{\sigma_n}$ as well as $e^{-V(\psi+\varphi)}$ also have the same form. Hence it makes sense to evaluate

$$S_{\sigma_1 \ldots \sigma_n}(x_1, \ldots, x_n)_{U,R,\Lambda,\beta} = \frac{\int P(d\psi) e^{-V} \psi_{x_1}^{\sigma_1} \ldots \psi_{x_n}^{\sigma_n}}{\int P(d\psi) e^{-V}} \qquad (A2.14)$$

or

$$\int P(d\psi) e^{-V(\psi+\varphi)} . \qquad (A2.15)$$

It follows from the above definitions and the Hadamard inequality that the numerator and denominator in (A2.14) and (A2.15) are entire functions of the variable λ. Assuming $\hat{v}(k) \geq 0$, it can be proved that the analyticity domain in λ is, uniformly in U, R, β for fixed Λ, the right half plane in general, and the full plane if $d = 1$ or the full half plane plus a disk around the origin if $d = 2$. In fact, if $d = 1$ the Hamiltonian is stable for all λ's, while if $d = 2$ it is stable only if λ is small or if $\mathcal{R}] \lambda > 0$.

It can be proved that the analyticity property is uniform in U, R, β if z is small enough and Λ is prefixed and that the limit as $U \to \infty, R \to -\infty, \beta \to \infty$ coincides with the limit as $\beta \to \infty$ of the ratio of traces in (3.2) (which in fact is equal to the limit as $U \to \infty, R \to -\infty$ of $S_{U,R,\beta,\Lambda}$ in (A2.14)). Hence statement (3.6) follows.

Appendix 3
Trees and Feynman Graphs

We consider the definition of the effective potential (10.1) and make the change of coordinates $\psi^{(\leq 0)} + \varphi_0 = \psi$, getting

$$e^{-V_{\text{eff}}(\sqrt{Z_0}\,\varphi_0)} = c \int P_{Z_0}^{(0)}(d\psi^{(\leq 0)})e^{-V^{(0)}(\sqrt{Z_0}\,\psi^{(\leq 0)}+\varphi_0)} =$$

$$= e^{-\frac{1}{2}(\varphi_0, Z_0\Gamma_0^{-1}p^2\varphi_0)}\,c \cdot \qquad (A3.1)$$

$$\cdot \int P_{Z_0}^{(0)}(d\psi)e^{-V^{(0)}(\sqrt{Z_0}\,\psi)}e^{(\psi, Z_0\Gamma_0^{-1}p^2\varphi_0)} \,,$$

where c denotes the different (formal) normalizations. Hence,

$$V_{\text{eff}}(\varphi_0) = \frac{1}{2}Z_0(\varphi_0, \Gamma_0^{-1}p^2\varphi_0) - q^{\leq 0}(Z_0\Gamma_0^{-1}p^2\varphi_0) \,, \qquad (A3.2)$$

where $q^{\leq 0}(\varphi_0)$, defined by

$$e^{q^{(\leq 0)}(\varphi_0)} = \int P_{Z_0}^{(0)}(d\psi)e^{-V^{(0)}(\sqrt{Z_0}\,\psi)+(\psi,\varphi_0)} \,, \qquad (A3.3)$$

is the generating functional of the Schwinger functions. In particular the two-point Schwinger function can be calculated by

$$S^{(\leq 0)}(x - y) = \frac{\delta^2 q^{(\leq 0)}(\varphi)}{\delta\varphi_x \delta\varphi_y}\bigg|_{\varphi=0} \,. \qquad (A3.4)$$

We first remark that the evaluation of the anomalous effective potentials admit a graphical representation very similar to the one discussed in chapter 8 for the normal effective potential.

We represent $-V^{(0)}(\sqrt{Z_0}\,\psi)$ as

$$\begin{array}{c}\bullet\!\!\!\rule[0.5ex]{3cm}{0.4pt} \\ {\scriptstyle 0}\end{array} \,, \qquad (A3.5)$$

and we note that $\mathcal{L}_3 V^{(0)} = 0$ because no term proportional to $F_3(\sqrt{Z_0}\,\psi)$ is present in the potential (7.1).

Then we compute $-\tilde{V}^{(-1)}$ as

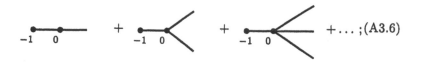

$$+\ldots\,; \quad (A3.6)$$

this differs from $V^{(-1)}$ of chapter 10 because each term, in general, will have a nonzero projection on $F_3(\psi) = \int (\partial \psi_x)^2\, dx$. The vertex labeled 0 represents integration with respect to a field with propagator $\frac{1}{Z_0}\tilde{g}^{(0)}$.

We apply to each term the operation $(1 - \mathcal{L}_3)$, collecting all the contributions to F_3 (obtained by applying $\mathcal{L}_3$ to each term) into a functional $\tilde{\alpha}_{-1} Z_0 F_3(\psi)$, and for later use we denote the coefficient of F_3 in $\mathcal{L}_3 V$ as $[\mathcal{L}_3 V]$. And we discard all such terms, which leave (as only trace of their coming to attention) the definition of Z_{-1}

$$Z_{-1} = Z_0 + 2\tilde{\alpha}_{-1} Z_0 = Z_0 + 2[\mathcal{L}_3 \tilde{V}^{(-1)}(\sqrt{Z_0}\,\psi)]\,. \qquad (A3.7)$$

The part obtained by applying $(1 - \mathcal{L}_3)$ will naturally represent a function of $\sqrt{Z_0}\,\psi$; however, we want to think of it as a function of $\sqrt{Z_{-1}}\,\psi$: this means that we imagine the n field kernels defining $V^{(-1)}(\sqrt{Z_{-1}}\,\psi)$ to be given by those of $(1 - \mathcal{L}_3)\tilde{V}^{(-1)}(\psi)$ multiplied by $\sqrt{\frac{Z_0}{Z_{-1}}}^n$.

The latter operation can be conveniently described in terms of an operator $\mathcal{O}_0$ which acts on a field monomial of degree n as

$$\mathcal{O}_h \psi_{x_1} \ldots \psi_{x_n} = \left(\sqrt{\frac{Z_h}{Z_{h-1}}}\right)^n \psi_{x_1} \ldots \psi_{x_n}\,, \qquad (A3.8)$$

with $h = 0$ (the more general case $h < 0$ will appear later).

The function $V^{(-1)}(\sqrt{Z_{-1}}\,\psi)$ is then split into its relevant part and its irrelevant part by the usual localization procedure,

$$V^{(-1)}(\sqrt{Z_{-1}}\,\psi) = \mathcal{L}V^{(-1)}(\sqrt{Z_{-1}}\,\psi) + (1 - \mathcal{L})V^{(-1)}(\sqrt{Z_{-1}}\,\psi) \quad (A3.9)$$

and

$$\mathcal{L}V^{(-1)} = \lambda_{-1}^{(6)} \int \sqrt{Z_{-1}}^6 \psi_x^6\, dx + 2\lambda_{-1}^{(4)} \int \sqrt{Z_{-1}}^4 \psi_x^4\, dx +$$
$$+ 2^2 \lambda_{-1}^{(2)} \int \sqrt{Z_{-1}}^2 \psi_x^2\, dx\,, \qquad (A3.10)$$

with no term proportional to $F_3 = \int (\partial \psi_x)^2\, dx$, by our construction.

We represent the splitting in (A3.9) graphically as

$$+ \ldots, (A3.11)$$

with

$$+ \ldots, (A3.12)$$

where the tilde over R, L is to remind us that we are following the anomalous integration procedure. The difference with respect to chapter 8, i.e., normal procedure, is "very slight": namely, the localization is $\mathcal{L}(1 - \mathcal{L}_3)$ for the local part and the remainder is $(1 - \mathcal{L})$, instead of the normal $\mathcal{L}$ and $(1 - \mathcal{L})$; furthermore the truncated expectations are evaluated with the propagator $\frac{1}{Z_0}\tilde{g}^{(0)}$ rather than $g^{(0)}$ (which, however, coincide accidentally as $Z_0 = 1$), and finally the n field kernels are altered by applying the operator $\mathcal{O}_0$ and are evaluated at $\sqrt{Z_{-1}}\,\psi$.

Hence the relevant and irrelevant parts of $V^{(-1)}(\psi)$ and the definition of Z_{-1} are simply

$$V_{rel}^{(-1)}(\sqrt{Z_{-1}}\,\psi) = \mathcal{O}_0 \mathcal{L}\,(1 - \mathcal{L}_3)\tilde{V}^{(-1)}(\sqrt{Z_0}\,\psi)\,,$$
$$V_{irr}^{(-1)}(\sqrt{Z_{-1}}\,\psi) = \mathcal{O}_0\,(1 - \mathcal{L})\tilde{V}^{(-1)}(\sqrt{Z_0}\,\psi)\,,$$
$$1 = \frac{Z_0}{Z_{-1}}\Big(1 + 2\big[\mathcal{L}_3\tilde{V}^{(-1)}(\sqrt{Z_0}\,\psi)\big]\Big)\,. \qquad (A3.13)$$

The iteration of the above calculation leads to the representation

$$V^{(h)}(\sqrt{Z_h}\,\psi) = V_{rel}^{(h)}(\sqrt{Z_h}\,\psi) + V_{irr}^{(h)}(\sqrt{Z_h}\,\psi) \qquad (A3.14)$$

and

$$V_{irr}^{(h)}(\psi) \;=\; \sum_{trees}$$ 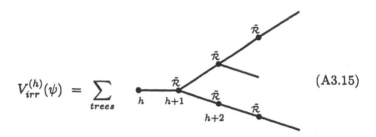 (A3.15)

where the tree vertices on scale k represent truncated expectations evaluated with the propagator $\frac{1}{Z_k}\tilde{g}^{(k)}$; also,

$$V_{rel}^{(h)}(\psi) \;=\; \underset{h\quad h+1}{\bullet\!\!-\!\!\bullet} \;+\; \sum_{trees}$$ 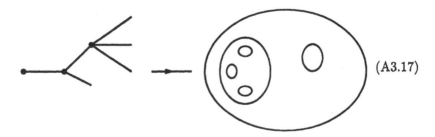 (A3.16)

It is now possible to describe the operations to be performed in the actual computations of the effective potential in terms of Feynman graphs.

The first step is to represent a tree as a hierarchy of boxes: each vertex v (including the trivial vertices and the endpoints) is represented by a box and all the $s_v \geq 1$ tree vertices $v_1, \ldots, v_{s_v}$ following it are represented as boxes drawn inside it, etc.

$$\text{(figure)} \quad\longrightarrow\quad \text{(figure)} \qquad\qquad \text{(A3.17)}$$

Each tree vertex carries in our constructions a scale label corresponding to the scale of the propagator to be used in the evaluation of the truncated expectation that it represents. If p is such label, then, in the tree representation, it will be attached to the corresponding box.

Inside the innermost boxes one draws the "graph elements" representing the relevant terms of the theory. In the φ^4 case in $d = 3$ one has

$$(A3.18)$$

Such graph elements represent, respectively,

$$\sqrt{Z_p}\,\lambda_p^{(j)}\psi_x^j, \qquad j = 6, 4, 2\,, \qquad (A3.19)$$

if p is the scale label of the considered box.

Then we form a graph by joining together some of the lines of the graph elements; the pairs of lines thus joined form *an inner line* and the ones that are not paired with others are called *external lines*. Furthermore,

1. Only n lines are regarded as external. And all the internal lines are drawn in such a way to be entirely contained at least in one of the boxes.
2. Each inner line is assigned a scale identical to the scale of the smallest box enclosing it.
3. Each box must contain a connected graph formed by its inner lines.
4. Each box must contain at least one inner line that is not contained in smaller boxes, if the corresponding tree vertex is not trivial, that is s_v is greater than 1.

To each such graph (obviously called a Feynman graph) G,

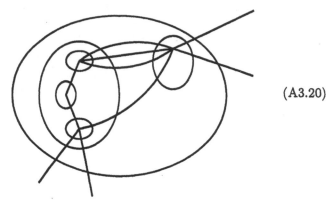

$$(A3.20)$$

we associate a kernel,

$$V^G \begin{pmatrix} x_1, \ldots, x_m \\ \nu_1, \ldots, \nu_m \end{pmatrix}, \qquad (A3.21)$$

which is defined by first multiplying the following factors:

a. A factor $2^{(d-\delta j)k_v} \lambda_{k_v}^{(j)}$ for each graph element of type j attached to a point in a box v with scale label k_v (with $\delta = (d-2)/2 = 1/2$).

b. A factor $(Z_{k_1} Z_{k_2}/Z_k^2)^{1/2}$ for each inner line of scale k joining two space vertices x, y of scales k_1, k_2, the scale of a space vertex being the one of the innermost box containing it.

c. A factor $(Z_k/Z_h)^{1/2}$ for each external line emerging from a space vertex of scale k.

d. A factor $\tilde{g}^{(k)}(x-y)$ for each inner line of scale k joining the vertices x, y.

The contribution to the effective potential from the selected graph is then obtained by integrating over all the positions of the space vertices the just-built kernel times a product of fields ψ_{x_i}, one per each external line emerging from the vertex x_i. Such contribution will have the form

$$\int V^G \begin{pmatrix} x_1, \ldots, x_p \\ \nu_1, \ldots, \nu_p \end{pmatrix} \prod_{j=1}^{p} \psi_{x_j}^{\nu_j} \qquad (A3.22)$$

if we assume that the points out of which the external lines emerge are the first q.

However, the above simple rule has to be modified if the graph has $q = 2, 4, 6$ external lines or even if its subgraphs associated with the inner boxes present $2, 4, 6$ external lines.

In the last cases a second operation has to be performed: it describes graphically the operations of subtraction that are implicit in the $\mathcal{L}$, $1-\mathcal{L}$, $\mathcal{L}_3$ operations. One looks at the boxes, called *renormalization boxes* below, containing subgraphs with $2,4,6$ lines (inner or external) going out of them.

The simplest situation arises when the boxes with the above property are disjoint (*nonoverlapping*). In such case, consider one of the renormalization boxes. Associated with it there will be a product:

$$C(\vec{x}, \vec{x}') = \prod_{j=1}^{q'} \tilde{g}^{(\bar{h}_j)}(x_j - x_{j'}) \prod_{j=q'+1}^{q} \psi_{x_j}, \qquad (A3.23)$$

where $\vec{x} = (x_1, \ldots, x_q)$ are the vertices inside the box and $\vec{x}' = (x_1', \ldots, x_{q'}')$ are the points where the q' inner lines that are external to the box end. Then the factor (A3.23) appearing in (A3.21) has to be replaced in (A3.21) by

$$\Delta C(\vec{x}, \vec{x}') = C(\vec{x}, \vec{x}') - \text{Taylor expansion in } \vec{x} - \vec{x}_1 \text{ of } C(\vec{x}, \vec{x}')$$

$$\text{to order 2 if } q = 2 \text{ or to order 0 otherwise}, \qquad (A3.24)$$

where $\vec{x}_1$ is the q-ple $(x_1, x_1, \ldots, x_1)$.

The equation (A3.24) is not obviously generalizable to the case when the renormalization box contains other renormalization boxes; this unpleasant problem was lurking in the literature until it was solved by Hepp ([H]), and it is still lurking because the latter paper is not as well known as it deserves to be.

The description, following the work of Gallavotti and Nicoló (see [G1]), is somewhat technical. One needs a table of the results of the localization operator

$$\mathcal{L} : \varphi_{x_1} \ldots \varphi_{x_6} : = : \varphi_{x_1}^6 :$$

$$\mathcal{L} : \varphi_{x_1} \ldots \varphi_{x_4} : = : \varphi_{x_1}^4 : + \left[\sum_{j=2}^{4} \varphi_{x_1}^3 (x_j - x_1) \partial \varphi_{x_1} \right] \qquad (A3.25)$$

$$\mathcal{L} : \varphi_{x_1} \varphi_{x_2} : = : \varphi_{x_1} \left(\varphi_{x_1} + [(x_2 - x_1)\partial \varphi_{x_1}] + \frac{1}{2}(x_2 - x_1)^2 \partial^2 \varphi_{x_1} \right) : ,$$

where the terms in square brackets can be omitted as the above expressions always appear multiplied by a kernel, symmetric function of the field arguments; after integration the terms in square brackets are formally integrals of exact derivatives and give zero "after integration by parts." Here there is, in fact, a further problem to worry about because the last statement is strictly correct only if one is considering a finite box with periodic boundary conditions: in this case, however, $(x_j - x_i)$ does not really make sense, being nonperiodic, and it should be replaced by $(x_j)_\alpha - (x_i)_\alpha \to L \sin L^{-1}((x_j)_\alpha - (x_i)_\alpha)$, if L is the side size of the periodic box. We, however, ignore this boundary condition problem, contenting ourselves at having hinted at its solution (see [G1], chapter 18).

The operation $1 - \mathcal{L}$ can be conveniently described in terms of the following fields

$$
\begin{aligned}
D_{x_2 x_1} &= \psi_{x_2} - \psi_{x_1} \,, \\
S_{x_2 x_1} &= \psi_{x_2} - \psi_{x_1} - (x_2 - x_1)\partial\psi_{x_1} \,, \\
T_{x_2 x_1} &= \psi_{x_2} - \psi_{x_1} - (x_2 - x_1)\partial\psi_{x_1} - \frac{1}{2}(x_2 - x_1)^2\partial^2\psi_{x_1} \,, \\
\overline{S}_{x_3 x_2 x_1} &= (x_3 - x_2)(\partial\psi_{x_2} - \partial\psi_{x_1}) \,.
\end{aligned}
\tag{A3.26}
$$

Then one can show inductively that $V_{irr}^{(h)}$ will admit a representation,

$$
\sum_{g>0} \int dx\, du\, ds\, dt\, d\bar{t}
\tag{A3.27}
$$

$$
V\left(
\begin{array}{ccccc}
\overbrace{x_1, ..}^{a} & \overbrace{(u_1, u_1'), ..}^{b} & \overbrace{(s_1, s_1'), ..}^{c} & \overbrace{(t_1, t_1'), ..}^{f} & \overbrace{(\bar{t}_1, \bar{t}_1', \bar{t}_1''), ..}^{\bar{f}} \\
\xi_1, .. & \alpha_1, \quad .. & \sigma_1, \quad .. & \tau_1, \quad .. & \bar{\tau}_1, \quad ..
\end{array}
\right)
$$

$$
\sqrt{Z_h}^{\,n+n'} \prod^{a}(\psi_{x_i})^{\xi_i} \prod^{b}(D_{u_i u_i'})^{\alpha_i} \prod^{c}(S_{s_i s_i'})^{\sigma_i} \prod^{t}(T_{t_i t_i'})^{\tau_i} \prod^{\bar{t}}(\overline{S}_{\bar{t}_i, \bar{t}_i', \bar{t}_i''})^{\bar{\tau}_i} \,,
$$

where $a, b, c, f, \bar{f}$ are the number of terms under the curly brackets and $x, u, s, t, \bar{t}$ are points in $\mathbf{R}^d$, and $\xi, \alpha, \sigma, \tau, \bar{\tau}$ are integers ≤ 4, and

$$
\begin{aligned}
m &= a + 2b + 2c + 2f + 3\bar{f} \,, \\
\rho &= \sum \alpha_i + 2\sum \sigma_i + 3\sum \tau_i + 2\sum \bar{\tau}_i \,, \\
n &= \sum \xi_i + \sum \alpha_i + \sum \sigma_i + \sum \tau_i, \qquad n' = \sum \bar{\tau}_i \,, \\
g &= -dm_0 + d + \rho + \frac{d-2}{2}n + \frac{d}{2}n' \,,
\end{aligned}
\tag{A3.28}
$$

$m_0 \leq m$ being the number of different space vertices appearing in the different fields.

The sum in (A3.27) runs over $g > 0$ and over various choices of the various labels. The kernels V that can appear in the sum are describable by the same Feynman graphs introduced above (see (A3.20)) with a few more labels attached.

More precisely, given a graph G of the type considered above, one looks at the innermost renormalization boxes and puts labels on their outcoming lines which have the interpretation that the fields they represent are no longer ψ fields but could also be $D, S, T, \overline{S}$ fields with arguments from the x's that are among the points out of which the external lines emerge.

The precise rule for attaching the labels is simply that one imagines applying the $\mathcal{R}$ operation to the product of fields represented by the renormalization box and the graph that it contains: such operation produces a linear combination of products of fields of the type in (A3.26); the labels affixed to the lines determine which term is actually selected.

The operation is repeated for the renormalization boxes containing renormalization boxes, the novelty being that now the lines that come out of a renormalization box may contain some nonlocal fields of type (A3.26) (produced by the action of $\mathcal{R}$ on the lines of some inner box). One just checks that the action of $\mathcal{R}$ can still be represented via the fields in (A3.26) and does not produce further nonlocal fields (in fact, for the innermost renormalization boxes one does not need all the fields in (A3.26)).

After finishing the "decoration" of the graph with all the labels (one can see that each graph cannot generate more than 4^{6n} labeled graphs), one evaluates the value of the graph by the rule of replacing each inner line by a propagator; but some of the inner lines may represent the new nonlocal fields, and the propagator corresponding to them has to be the propagator between the fields corresponding to the two lines that are contracted to form the inner line (such a propagator is just the integral of the product of the fields with respect to the measure $Z_k P_{Z_k}^{(k)}(d\psi)$).

In the end the total effective potential can be represented as a linear combination of not more than about $6^{6n} n!^3$ terms, each corresponding to a decorated graph evaluated as described.

If the construction of the effective potential is performed according to the above rules, it is quite simple to check that the bound (10.24) holds. For more details, see the analogous calculation described in [G1] for the $d = 4$ (normal scaling) theory: note that in the latter case φ^6 is an irrelevant term so that the bounds improve (as the number of Feynman graphs of order n grows as $n!^2$ rather than $n!^3$ and $n!^2$ is replaced by $n!$).

Appendix 4
Schwinger Functions and Anomalous Dimension

We compute $V_{\text{eff}}(\frac{1}{Z_0 p^2}\Gamma_0(p)\varphi_0)$ or, better, $q^{(\leq 0)}(\varphi_0)$, which, by (4.3) and (A3.2), (A3.3), is the generating functional of the truncated Schwinger functions,

$$e^{q^{(\leq 0)}(\varphi_0)} = \int P_{Z_0}^{(0)}(d\psi) e^{-V^{(0)}(\sqrt{Z_0}\,\psi)+(\varphi_0,\psi)} . \qquad (A4.1)$$

To compute the pair Schwinger function we evaluate the integral by using the procedure of chapter 10.

We have to evaluate only the part of $q^{(\leq 0)}$ that is quadratic in φ. We represent the argument of the exponential in (A4.1) inside the integral as

$$(A4.2)$$

and we suppose inductively that, fixing $Q_0 = 1, G_0 = 0$, the successive integrals over the various scales modify the exponent in the the integrand exponential into a functional representable as

$$(A4.3)$$

where the first term symbolizes the (anomalous) effective potential constructed in chapter 10, the second and the third terms represent (for suitable convolution operators Q_h, G_h), respectively,

$$\int dx\, \varphi_x\, (Q_h * \psi)_x , \qquad (A4.4)$$

and the functional

$$\int dx\, \varphi_x\, [G_h * \frac{\delta V^{(h)}}{\delta\psi}]_x , \qquad (A4.5)$$

and, finally, the last term represents the contribution of all the connected graphs with two φ fields produced by the successive integrations.

Then, integrating as usual with the method of chapters 8 and 10, we can express the result of the integral over the momentum shell of scale h graphically as

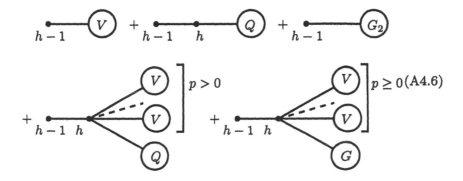

We see that the sum of the graphs in the first term of the second line is

$$Q_h * \frac{1}{Z_h} \tilde{g}^{(h)} * \frac{\delta \bar{V}^{(h-1)}}{\delta \psi} , \qquad (A4.7)$$

and the sum of the graphs in the last term is

$$G_h * \frac{\delta \bar{V}^{(h-1)}}{\delta \psi} . \qquad (A4.8)$$

So that, writing $\bar{V}^{(h-1)}(\sqrt{Z_h}\,\psi) = V^{(h-1)}(\sqrt{Z_{h-1}}\,\psi) + \frac{1}{2} z_{h-1} Z_h F_3(\psi)$, we see that

$$Q_{h-1} = Q_h - z_{h-1} \sum_{j=h}^{0} \frac{Z_h}{Z_j} w_j * Q_j, \qquad Q_0 = 1 ,$$

$$\qquad\qquad (A4.9)$$

$$G_{h-1} = \sum_{j=h}^{0} \frac{1}{Z_j} \tilde{g}^{(j)} * Q_j, \qquad w_h(k) = k^2 \tilde{g}^{(h)}(k) ,$$

and we see that the second-order term can be simply evaluated by a graphical rule almost identical to the one elaborated in chapter 10 and appendix 3, but with the following modifications (see fig. (A4.10)):

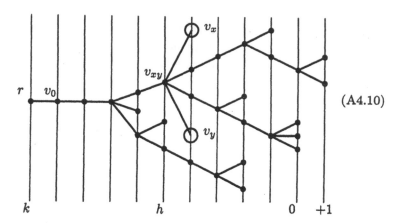

$$(A4.10)$$

1. There are $n + 2$ endpoints, $n \geq 0$, and two of them, denoted v_x and v_y in the figure, represent the following function

$$\int dx \, \varphi_x \left[Q_h * \psi_x^{(\leq h)} + G_h * \frac{\delta}{\delta \psi_x} \left(V^{(h)} (\sqrt{Z_h} \psi) \right) \right] , \qquad (A4.11)$$

where the following recursive relations for the convolution operators Q_h, G_h hold:

$$Q_{h-1} = Q_h - z_{h-1} Z_h [-\partial^2] G_{h-1} = Q_h - z_{h-1} \sum_{j=h}^{0} \frac{Z_h}{Z_j} w_j * Q_j ,$$

$$G_{h-1} = G_h + \tilde{g}^{(h)} * Q_h = \sum_{j=h}^{0} \frac{1}{Z_j} \tilde{g}^{(j)} * Q_j , \qquad (A4.12)$$

$$w_h(k) = k^2 \, \tilde{g}^{(h)}(k) .$$

2. The two special endpoints of item (1) belong to the vertical line with frequency index $h+1$ and are attached at the *same* tree vertex v_{xy} bearing a frequency label h. This implies that h is the scale at which the lines φ_x and φ_y become connected by graph lines.

3. There are no external lines in the root of the tree.

4. There are no $\mathcal{R}^*$ labels associated with the tree vertices v belonging to the line l joining the root to v_{xy}.

It can be checked, by the same techniques of [BGPS], that $\tilde{g}^{(h)}(x)$ satisfies a bound such as

$$|\tilde{g}^{(h)} * Q_h(x)| \leq C \gamma^h e^{-\kappa \gamma^h |x|} \qquad (A4.13)$$

for some $\kappa > 0$, provided the z_h verify $|z_h| \leq C\varepsilon^2$ for all h, with ε small enough (i.e., by the bounds of chapter 5, provided the running couplings r_h verify $|r_h| < \varepsilon$ for ε small enough).

Hence, for the purpose of establishing bounds in x space, we could replace Q_h by 1. It is possible to prove that a similar property is valid in k space, the key property being the consequence of the recursive definition of Q_h,

$$Q_h(k) = \left[1 + \varepsilon^2 \, b_h(\gamma^{-h}k) \, \log \left(e + (\frac{k\gamma^{-h}}{p_0})^2\right)\right] \cdot$$
$$\cdot \, (1 + (\frac{k\gamma^{-h}}{p_0})^2)^{C\varepsilon^2} \, , \tag{A4.14}$$

where $C > 0$ is a constant and $b_h(k)$ is analytic in a strip of imaginary width $p_0\kappa$ (for some $\kappa > 0$) around the real axis (in the two k variables) and is bounded there by a constant B (h independent), provided the z_h verify $|z_h| \leq \varepsilon^2$ for all h with ε small enough (i.e., by the discussion of chapter 10 we expect that ε is of the order of $\max_h |v_h|^2$).

The latter (A4.14) shows that, for the purpose of establishing bounds (both in x space and in k space), we could replace in (6.14) Q_h by 1 and G_h by $\hat{g}^{(h)} = \sum_{h'>h} \frac{1}{Z_h} \tilde{g}^{(h')}$.

The (A4.13), or (A4.14), implies (see [BGPS]) that

Theorem

The pair Schwinger function can be written in the form

$$S_{int}(x - y) = \sum_{h=-\infty}^{0} \frac{1}{Z_h} \tilde{g}^{(h)}(x - y)\,(1 + O(\varepsilon)) \, , \tag{A4.15}$$

where ε is supposed to be small enough and to be a bound on the running couplings on all scales; and

$$|\tilde{g}^{(h)}(x - y)| \leq B\gamma^h e^{-\bar{\kappa}|x-y|} \, , \tag{A4.16}$$

$B > 0, \bar{\kappa}$ being suitable constants, independent on ε. The $O(\varepsilon)$ is a function, which is a formal power series in the running couplings and with coefficients that are bounded by $n!^2$ to order n, independently on $x - y$.

An immediate corollary of the theorem is that the pair Schwinger function decays for $|x - y| \to \infty$ to all orders of perturbation expansion in the running couplings, as $|x - y|^{-1-2\eta}$, with η defined by
$$Z_h/Z_{h+1} \xrightarrow[h\to-\infty]{} 2^{2\eta}.$$

Appendix 5
Propagators for the Bose Gas

Let $P^{(t)}(d\psi)$ be the formal complex Gaussian measure with covariance

$$g^{(t)}(x) = \frac{1}{(2\pi)^4} \int dk \, e^{-ikx} \frac{t(k)}{-ik_0 + \frac{\vec{k}^2}{2m}} , \qquad (A5.1)$$

where $x = (x_0, \vec{x})$, $k = (k_0, \vec{k})$ and $t(k)$ is a positive cutoff function.

We consider the fields $\chi_x^\pm = \frac{1}{\sqrt{2\rho}}(\psi_x^+ \pm \psi_x^-) = (2\pi)^{-2} \int dk e^{\pm ikx} \chi_k^\pm$. Their propagator has the form

$$\langle \chi_x^\sigma \chi_y^{\sigma'} \rangle = \frac{1}{(2\pi)^4} \int e^{-ikx} t(k) G^{-1}(k)_{\sigma\sigma'} , \qquad (A5.2)$$

where the matrix G, which we call the propagator matrix, is

$$G = \rho \begin{pmatrix} \frac{\vec{k}^2}{2m} & ik_0 \\ -ik_0 & -\frac{\vec{k}^2}{2m} \end{pmatrix} . \qquad (A5.3)$$

Therefore the formal Gaussian measure,

$$P^{(t)}(d\psi) e^{-\rho \sum_{\sigma\sigma'} \int \chi_{\sigma k}^\sigma \Delta_{\sigma\sigma'} \chi_{-\sigma' k}^{\sigma'} dk} , \qquad (A5.4)$$

has a propagator matrix $G' \equiv G + 2\rho\Delta t(k)$. In particular, if Δ is associated with the quadratic form

$$\int dx [2a(\chi_x^+)^2 - 2\sum_{i=0}^{3} b_i(\partial_i \chi_x^-)^2 - 2c\chi_x^+ \partial_0 \chi_x^-] , \qquad (A5.5)$$

with $a, b_0, b = b_1 = b_2 = b_3, c$ not negative real numbers, that is,

$$\Delta = \begin{pmatrix} 2a & ick_0 \\ -ick_0 & -2(b_0 k_0^2 + b\vec{k}^2) \end{pmatrix} , \qquad (A5.6)$$

then it is

$$G' = \rho \begin{pmatrix} \frac{\vec{k}^2}{2m} + 4at(k) & ik_0[1 + 2ct(k)] \\ -ik_0[1 + 2ct(k)] & -\frac{\vec{k}^2}{2m} - 4(b k_0^2 + b\vec{k}^2)t(k) \end{pmatrix} . \qquad (A5.7)$$

And the $\psi^{\pm}$ field propagators with respect to the measure $P_{ab}^{(t)}(d\psi)$ with propagator matrix G' can be immediately checked to be

$$\int P_{a,b}^{(t)}(d\psi)\psi_x^- \psi_y^+ = \frac{1}{(2\pi)^4}\int dk\, e^{-ik(x-y)}.$$ (5.8)

$$\cdot \frac{ik_0[1 + 2ct(k)] + \vec{k}^2 + 2d^+(k)t(k)}{[1 + 2ct(k)]^2 k_0^2 + \vec{k}^4 + 4d^+(k)\vec{k}^2 t(k) + 16a(\sum_{i=0}^3 b_i k_i^2)t(k)^2}\, t(k)$$

and

$$\int P_{a,b}^{(t)}(d\psi)\psi_x^+ \psi_y^+ = \int P_{a,b}^{(t)}(d\psi)\psi_x^- \psi_y^- =$$

$$= \frac{1}{(2\pi)^4}\int dk\, e^{-ik(x-y)}.$$ (5.9)

$$\cdot \frac{-2d^-(k)t(k)}{[1 + 2ct(k)]^2 k_0^2 + \vec{k}^4 + 4d^+(k)\vec{k}^2 t(k) + 16a(\sum_{i=0}^3 b_i k_i^2)t(k)^2}\, t(k)\,,$$

where $d^{\pm}(k) = a \pm (b_0 k_0^2 + b\vec{k}^2)$. Note that (5.9) can be derived also by performing a Bogoliubov transformation.

Appendix 6
The Beta Function for the Bose Gas

In this appendix we want to prove (15.18). We first note that all the terms in the r.h.s. of (15.18) are obtained by applying the localization operator to

$$V_h^{\leq 4} \equiv \mathcal{E}_h(V_h) - \frac{1}{2!}\mathcal{E}_h^T(V_h^2) + \frac{1}{3!}\mathcal{E}_h^T(V_h^3) - \frac{1}{4!}\mathcal{E}_h^T(V_h^4) , \qquad (A6.1)$$

where $\mathcal{E}_h$ and $\mathcal{E}_h^T$ denote, respectively, the expectation and the truncated expectation with respect to the measure describing the fluctuations on scale h, whose propagator is given by

$$\gamma^{(3+\frac{\sigma_1+\sigma_2}{2})h}\tilde{g}_h^{\sigma_1\sigma_2}(\gamma^h x) , \qquad (A6.2)$$

where

$$\tilde{g}_h^{\sigma_1\sigma_2}(x) = \frac{1}{(2\pi)^4}\int dk \frac{e^{-ikx} T_0(k)\, \bar{p}_h^{\sigma_1\sigma_2}(k)}{\rho \mathcal{D}_h(k)} \qquad (A6.3)$$

$$\mathcal{D}_h(k) = b_h(k)k_0^2 + a_h(k)\frac{v_0^2}{4}\vec{k}^2 +$$

$$+ \gamma^{2h}[\frac{\vec{k}^4}{4m^2}(1 + 4A_h(k)) + 4B_h(k)\frac{k_0^2\vec{k}^2}{p_0^2}] , \qquad (A6.4)$$

$$a_h(k) = 4Z_h(k) + 16Z_h(k)A_h(k) ,$$
$$b_h(k) = E_h(k)^2 + 16Z_h(k)B_h(k) , \qquad (A6.5)$$
$$\bar{p}_h^{-+}(k) = \bar{p}_h^{+-}(-k) = -ik_0 E_h(k) ,$$
$$\bar{p}_h^{++}(k) = \frac{\vec{k}^2}{2m} + 4[B_h(k)k_0^2\frac{2m}{p_0^2} + A_h(k)\frac{\vec{k}^2}{2m}] ,$$
$$\bar{p}_h^{--}(k) = -\gamma^{2h}\frac{\vec{k}^2}{2m} - 4Z_h(k)\frac{p_0^2}{2m} . \qquad (A6.6)$$

Some other remarks are important.

1. In order to calculate the beta function one has to evaluate some Feynman graphs at zero momentum of the external lines. Therefore, only terms without internal lines or terms with at least one loop can contribute, since the single scale propagator vanishes at zero momentum.

2. The previous remark also implies that, in the graphs with only one loop, all internal lines must carry the same momentum. Hence, if we suitably choose the cutoff function $t_0(k)$, the internal lines of the loop may only have propagators of scale h or $h + 1$; in fact, at least one propagator must be of scale h and the supports of the Fourier transforms of the propagators of scale h and $h' \geq h$ are disjoint if $h' > h + 1$. This implies that the trees (see chapter 8 for the definitions) that one has to consider in evaluating the contribution of such graphs are the trees with only one vertex of scale h and at most three endpoints, and the trees with one vertex on scale h and one nontrivial vertex on scale $h + 1$, corresponding to some irrelevant contribution.

3. In the calculation of the graphs with only one loop, the subgraph associated with the tree vertex of scale $h + 1$ has no loop. Therefore, in this tree vertex the $\mathcal{R}$ operation coincides with the identity.

4. Since we are interested only in the leading orders, we can neglect in the rescaled propagators (A6.3) the terms proportional to γ^{2h} and the dependence on k of Z_h, A_h, B_h, and E_h (see (14.20)). For the same reason we can approximate Z_{h+1}, A_{h+1}, B_{h+1} and E_{h+1} by Z_h, A_h, B_h, and E_h in the expression of $\tilde{g}_{h+1}^{\sigma_1 \sigma_2}(x)$. Finally, we can neglect the difference between λ_{h+1}, μ_{h+1} and λ_h, μ_h in the endpoints of the trees involving a tree vertex on scale $h + 1$.

The previous remarks imply that the leading terms in the beta function can be obtained by the following steps.

1. Evaluate the graphs with one loop and propagator given by the sum of the single-scale propagators of scale h and $h + 1$, approximated as explained in remark (4), that is,

$$\gamma^{(3+\frac{1}{2}(\sigma_1+\sigma_2))h} g_{1,h}^{\sigma_1 \sigma_2}(\gamma^h x) , \qquad (A6.7)$$

where, if $\mathcal{D}_h(k)$ denotes the determinant of the propagator matrix $G_{(h)}$, see (14.7),

$$g_{1,h}^{\sigma_1 \sigma_2}(x) = \frac{1}{(2\pi)^4} \int dk \, e^{-ikx} \frac{T_1(k) \bar{p}_h^{\sigma_1 \sigma_2}(k)}{\rho \mathcal{D}_h(k)} . \qquad (A6.8)$$

Here $\mathcal{D}_h(k)$ and $\bar{p}_h^{\sigma_1 \sigma_2}$ denote the expressions (A6.4) and (A6.6), modified as explained in remark (4) above, that is,

$$\mathcal{D}_h(k) = b_h k_0^2 + a_h \frac{v_0^2}{4} \vec{k}^2 , \qquad (A6.9)$$

$$a_h = 4Z_h(1 + 4A_h), \qquad b_h = E_h^2 + 16Z_h B_h , \qquad (A6.10)$$

$$\bar{p}_h^{--}(k) = -4q_0 Z_h, \qquad\qquad \bar{p}_h^{-+}(k) = \bar{p}_h^{+-}(-k) = -ik_0 E_h ,$$

$$\bar{p}_h^{++}(k) = \frac{16 Z_h B_h k_0^2 + a_h \vec{k}^2 \frac{v_0^2}{4}}{4 q_0 Z_h}, \qquad\qquad q_0 = \frac{p_0^2}{2m} , \qquad\qquad (A6.11)$$

and

$$T_1(k) = T_0(k) + T_0(\gamma^{-1}k) = t_0(\gamma^{-1}k) - t_0(\gamma k) . \qquad\qquad (A6.12)$$

2. Evaluate the same graphs with propagator of scale $h + 1$, again approximated as in remark (4). This propagator is obtained from (A6.8) by substituting $T_1(k)$ with

$$T_2(k) = T_0(\gamma^{-1}k) = t_0(\gamma^{-1}k) - t_0(k) . \qquad\qquad (A6.13)$$

We shall denote $g_{2,h}^{\sigma_1\sigma_2}$ the corresponding rescaled propagator.

3. Subtract the values found in (2) from the values found in (1) and add the trivial graphs without any internal line.

4. Approximate in the result the cutoff function $t_0(k)$ by the characteristic function of the set $\{k_0^2 + \vec{k}^2 \leq 1\}$. Note that this approximation is everywhere equivalent to calculating the graphs with all propagators on the single scale h, except in the case of the beta function for B_{h-1}, A_{h-1}, and E_{h-1}, which involve derivatives with respect to the loop momentum. Hence, except in this case, we have to calculate the graphs by using the propagator obtained from (A6.8) by substituting $T_1(k)$ with $T_0(k)$; we shall denote $g_h^{\sigma_1\sigma_2}$ this rescaled propagator.

The trivial graphs give the linear terms in the r.h.s. of (15.18), except the term linear in λ_h appearing in the equation for ν_{h-1}. This term is obtained by contracting two χ^- fields in the λ_h vertex; one gets

$$-\binom{4}{2} \frac{p_0^2}{2m} \lambda_h \gamma^{2h} 4 q_0 Z_h \beta_{1,h} F_{20} , \qquad\qquad (A6.14)$$

where F_{20} is defined as in (14.10) and

$$\beta_{1,h} = \int \frac{dk}{(2\pi)^4} \frac{T_0(k)}{\mathcal{D}_h(k)} = \frac{1 - \gamma^{-2}}{8\pi^2(\sqrt{a_h b_h} + a_h)} q_0^2 \left(\frac{2}{v_0}\right)^3 . \qquad\qquad (A6.15)$$

The quadratic terms in the equations for λ_h and μ_h are associated with the graphs drawn in fig. (A6.16), where the heavy lines represent the χ^- fields and the dotted ones represent χ^+.

Note that the coefficients in front of the different graphs, here and in the following figures, indicate how many times the graph appears, if one expands the powers of the potential in the r.h.s. of (A6.1) in terms of the different monomials of the field, whose sum gives the potential, and consider the different possibilities of connecting different point vertices, giving rise to the same graph. In order to get the right contribution to the beta function, one has to consider also the coefficients of the expectations in (A6.1) and the combinatorial factors that count the different possibilities of choosing the external lines in the different vertices of the graph and the different possibilities of contracting the internal lines.

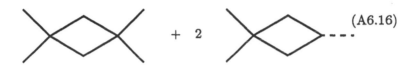

$$\tag{A6.16}$$

The contribution to $\mathcal{L}V_{h-1}(\chi)$ coming from the graphs in (A6.16) is

$$
\begin{aligned}
&-\frac{1}{2}\left[\binom{4}{2}^2 2\lambda_h^2(4q_0 Z_h)^2\beta_{2,h}F_{40}\right.\\
&\left.+2\binom{4}{2}2\lambda_h\mu_h(4q_0 Z_h)^2\beta_{2,h}F_{21}\right]\left(\frac{p_0^2}{2m}\right)^2 ,
\end{aligned}
\tag{A6.17}
$$

where

$$
\beta_{2,h}=\int\frac{dk}{(2\pi)^4}\frac{T_0(k)}{\mathcal{D}_h(k)^2}=\frac{\log\gamma}{8\pi^2\sqrt{a_h^3 b_h}}\left(\frac{2}{v_0}\right)^3 .
\tag{A6.18}
$$

The cubic terms in the equations for λ_h and μ_h are associated with the graphs drawn in (A6.19).

$$\tag{A6.19}$$

The contribution to $\mathcal{L}V_{h-1}(\chi)$ coming from the graphs in (A6.19) is

$$\frac{1}{6}\, 3\cdot 8\left[\binom{4}{2}\lambda_h\mu_h^2 F_{40} + \mu_h^3 F_{21}\right]\left(\frac{p_0^2}{2m}\right)^3 \beta_{3,h}\,, \qquad (A6.20)$$

where

$$\begin{aligned}
\beta_{3,h} &= \rho^3 \int \frac{dk}{(2\pi)^4}\left[g_h^{--}(k)^2 g_h^{++}(k) + 2g_h^{--}(k)g_h^{-+}(k)^2 +\right.\\
&\quad \left. + g_h^{--}(k)g_h^{-+}(k)g_h^{+-}(k)\right] =\\
&= \rho^3 \int \frac{dk}{(2\pi)^4}\left[g_h^{--}(k)^2 g_h^{++}(k) + g_h^{--}(k)g_h^{-+}(k)^2\right] =\\
&= 4q_0 Z_h \beta_{2,h}\,.
\end{aligned} \qquad (A6.21)$$

The last equality follows from the identity

$$\rho^2[g_h^{--}(k)g_h^{++}(k) + g_h^{-+}(k)^2] = -\frac{T_0(k)^2}{\mathcal{D}_h(k)} \qquad (A6.22)$$

and from the observation that $T_0(k)^2 = T_0(k)$, in the approximation of item (4) above. We also used the fact that $g_h^{-+}(k) = -g_h^{+-}(k) = -g_h^{-+}(-k)$. Hence (A6.20) can be written as

$$4\left[\binom{4}{2}\lambda_h\mu_h^2 F_{40} + \mu_h^3 F_{21}\right]4q_0 Z_h\left(\frac{p_0^2}{2m}\right)^3 \beta_{2,h}\,. \qquad (A6.23)$$

The quartic term in the equation for λ_h is associated with the graphs drawn in (A6.24).

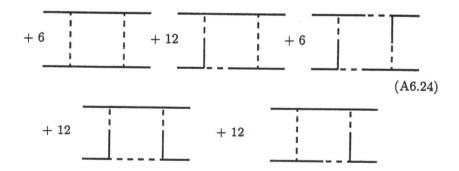

$$(A6.24)$$

The contribution to $\mathcal{L}V_{h-1}(\chi)$ coming from the graphs in (A6.24) is

$$-\frac{1}{4!} 2^4 \cdot 6 \left(\frac{p_0^2}{2m}\right)^4 \mu_h^4 \beta_{4,h} F_{40} , \qquad (A6.25)$$

where

$$\beta_{4,h} = \rho^4 \int \frac{dk}{(2\pi)^4} \Big[g_h^{--}(k)^2 g_h^{++}(k)^2 + 2g_h^{--}(k)g_h^{-+}(k)^2 g_h^{++}(k) +$$

$$+ g_h^{-+}(k)^4 - 2g_h^{--}(k)g_h^{-+}(k)^2 g_h^{++}(k) + 2g_h^{--}(k)g_h^{-+}(k)^2 g_h^{++}(k) \Big] =$$

$$= \rho^4 \int \frac{dk}{(2\pi)^4} \Big[g_h^{--}(k)g_h^{++}(k) + g_h^{-+}(k)^2 \Big]^2 = \beta_{2,h} , \qquad (A6.26)$$

where we used again the identity (A6.22).

The leading contribution in the flow equation for Z_h is the quadratic term associated with the graph in the first line of (A6.28), whose contribution to the local part is

$$-\frac{1}{2} 2 \left(\frac{p_0^2}{2m}\right)^2 \mu_h^2 (4q_0 Z_h)^2 \beta_{2,h} F_{02} \qquad (A6.27)$$

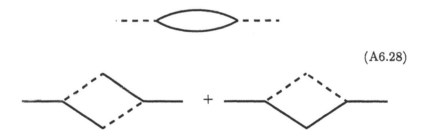

$$(A6.28)$$

and no operators involving field derivatives arise because this is a marginal operator.

The contribution to the local part of the last two graphs of (A6.28) is

$$-\frac{1}{2} 2^2 \left(\frac{p_0^2}{2m}\right)^2 \mu_h^2 \Big\{ \gamma^{2h} F_{20} \int \frac{dk}{(2\pi)^4} \Big[g_h^{--}(k)g_h^{++}(k) + g_h^{-+}(k)^2 \Big] -$$

$$-\frac{1}{2} [\beta_{t,h}^{(1)} - \beta_{t,h}^{(2)}] q_0^2 D_{tt} - \frac{1}{2} [\beta_{s,h}^{(1)} - \beta_{s,h}^{(2)}] 4m^2 q_0^2 D_{ss} \Big\} , \qquad (A6.29)$$

where D_{tt} and D_{ss} are defined as in (14.9) and

$$\beta_{t,h}^{(i)} = \int dx \, x_0^2 \Big[g_{i,h}^{-+}(x)^2 - g_{i,h}^{--}(x) g_{i,h}^{++}(x) \Big] \,,$$

$$\beta_{s,h}^{(i)} = \int dx \, \vec{x}^2 \Big[g_{i,h}^{-+}(x)^2 - g_{i,h}^{--}(x) g_{i,h}^{++}(x) \Big] \,, \tag{A6.30}$$

$$g_{i,h}^{\sigma_1 \sigma_2}(x) = \frac{1}{(2\pi)^4} \int dk \frac{e^{-ikx} T_i(k) \, \bar{p}_h^{\sigma_1 \sigma_2}(k)}{\rho \mathcal{D}_h(k)} \,. \tag{A6.31}$$

We have

$$\rho^2 \beta_{t,h}^{(i)} = \int \frac{dk}{(2\pi)^4} \Big\{ E_h^2 \Big[\frac{\partial}{\partial k_0} \frac{k_0 T_i}{\mathcal{D}_h} \Big]^2 +$$

$$+ \Big[\frac{\partial}{\partial k_0} \frac{T_i}{\mathcal{D}_h} \Big] \frac{\partial}{\partial k_0} \Big[\Big(1 - \frac{E_h^2 k_0^2}{\mathcal{D}_h} \Big) T_i \Big] \Big\} = \tag{A6.32}$$

$$= \int \frac{dk}{(2\pi)^4} \Big\{ E_h^2 \frac{T_i^2}{\mathcal{D}_h^2} + \Big(\frac{\partial}{\partial k_0} \frac{T_i}{\mathcal{D}_h} \Big) \frac{\partial T_i}{\partial k_0} \Big\} \,.$$

It is easy to see that the terms containing the derivatives of T_i give the same contribution to $\beta_{t,h}^{(1)}$ and $\beta_{t,h}^{(2)}$, so that, by doing in the remaining term the approximation of item (4) above, we get

$$\rho^2 [\beta_{t,h}^{(1)} - \beta_{t,h}^{(2)}] = \beta_{2,h} E_h^2 \,. \tag{A6.33}$$

We have also

$$\rho^2 \beta_{s,h}^{(i)} = \sum_{j=1}^{3} \int \frac{dk}{(2\pi)^4} \Big\{ E_h^2 \Big[\frac{\partial}{\partial k_j} \frac{k_0 T_i}{\mathcal{D}_h} \Big]^2 +$$

$$= + \Big[\frac{\partial}{\partial k_j} \frac{T_i}{\mathcal{D}_h} \Big] \frac{\partial}{\partial k_j} \Big[\Big(1 - \frac{E_h^2 k_0^2}{\mathcal{D}_h} \Big) T_i \Big] \Big\} = \tag{A6.34}$$

$$= \sum_{j=1}^{3} \int \frac{dk}{(2\pi)^4} \Big(\frac{\partial}{\partial k_j} \frac{T_i}{\mathcal{D}_h} \Big) \frac{\partial T_i}{\partial k_j} \,,$$

and one can see again that the terms containing the derivative of T_i give the same contribution to $\beta_{s,h}^{(1)}$ and $\beta_{s,h}^{(2)}$, so that

$$\beta_{s,h}^{(1)} - \beta_{s,h}^{(2)} = 0 \,. \tag{A6.35}$$

In summary, the contribution to the local part of the last two graphs of (A6.28) is

$$2\beta_{1,h} \Big(\frac{p_0^2}{2m} \Big)^2 \mu_h^2 \gamma^{2h} F_{20} + \beta_{2,h} E_h^2 \Big(\frac{p_0^2}{2m} \Big)^2 \mu_h^2 q_0^2 D_{tt} \,. \tag{A6.36}$$

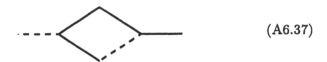

$$(A6.37)$$

The leading contribution in the flow equation for E_h is associated with the graph drawn in (A6.37), whose local part is calculated by taking the first-order Taylor expansion of the external χ^- field. The term of order zero, which would contribute to the relevant term F_{11}, cancels out for parity reasons, as well as the term of order one containing the spatial derivatives, so that the local part is equal to

$$-\frac{1}{2}\, 2\, 2\, 2 \left(\frac{p_0^2}{2m}\right)^2 \mu_h^2 (\beta_{5,h}^{(1)} - \beta_{5,h}^{(2)}) q_0(-D_t) \, , \qquad (A6.38)$$

where D_t is defined as in (14.9) and, if we use also the definition (A6.31),

$$\beta_{5,h}^{(i)} = \rho^2 \int dx \; x_0 g_{i,h}^{--}(x) g_{i,h}^{+-}(x) \, . \qquad (A6.39)$$

We have

$$\beta_{5,h}^{(i)} = 4 q_0 Z_h E_h \int \frac{dk}{(2\pi)^4} \frac{k_0 T_i}{D_h} \left[\frac{\partial}{\partial k_0} \frac{T_i}{D_h} \right] \, , \qquad (A6.40)$$

and we can see, as before, that the terms containing the derivatives of T_i give the same contribution to $\beta_{5,h}^{(1)}$ and $\beta_{5,h}^{(2)}$, so that, in the usual approximation,

$$\beta_{5,h}^{(1)} - \beta_{5,h}^{(2)} = -8 q_0 Z_h E_h b_h \int \frac{dk}{(2\pi)^4} \frac{k_0^2 T_0(k)}{D_h^3} = -2 q_0 Z_h E_h \beta_{2,h} \, , \quad (A6.41)$$

where the last equality follows from an explicit calculation and $\beta_{2,h}$ is exactly the same function of a_h and b_h defined in (A6.18). Hence, the local part of the graph drawn in (A6.37) can be written as

$$-8 \left(\frac{p_0^2}{2m}\right)^2 q_0^2 \mu_h^2 Z_h E_h \beta_{2,h} D_t \, . \qquad (A6.42)$$

The flow equations immediately follow from (A6.14), (A6.17), (A6.23), (A6.25), (A6.27), (A6.36), and (A6.42):

$$\lambda_{h-1} = \lambda_h - 36(4Z_h)^2 \beta_{2,h} \left[\lambda_h - \frac{\mu_h^2}{12 Z_h}\right]^2 \, ,$$

$$\mu_{h-1} = \mu_h - 12(4Z_h)^2 \beta_{2,h} \mu_h \left[\lambda_h - \frac{\mu_h^2}{12Z_h} \right] \ ,$$

$$Z_{h-1} = Z_h - 8Z_h^2 \beta_{2,h} \mu_h^2 \ ,$$

$$A_{h-1} = A_h \ ,$$

$$B_{h-1} = B_h + \frac{1}{2} \beta_{2,h} E_h^2 \mu_h^2 \ ,$$

$$E_{h-1} = E_h - 8Z_h E_h \beta_{2,h} \mu_h^2 \ ,$$

$$\nu_{h-1} = \gamma^2 \left[\nu_h + 24Z_h \beta_{1,h} \lambda_h - 2\beta_{1,h} \mu_h^2 \right] \ ,$$

$$(A6.43)$$

where

$$\beta_{2,h} = \frac{\log \gamma}{8\pi^2 \sqrt{a_h^3 b_h}} \frac{p_0^3}{\rho} , \qquad \beta_{1,h} = \frac{1 - \gamma^{-2}}{8\pi^2 (\sqrt{a_h b_h} + a_h)} \frac{p_0^3}{\rho} , \qquad (A6.44)$$

with

$$a_h = 4Z_h(1 + 4A_h), \qquad b_h = E_h^2 + 16Z_h B_h \ , \qquad (A6.45)$$

and the (15.18) follow immediately from the above by replacing μ_h via (15.7) and by adopting the notations declared for (15.18).

One may wonder whether it might be that the equation for μ_h is compatible with the others. Of course, if (15.7) is valid, it must be; this can also be seen directly (and in fact it is this remark that would make one conjecture the exact relation (15.7), if one did not know it). However, the compatibility is true only to the order of the calculation that we are performing, i.e., within the neglected corrections; if we did not know a priorithe validity of (15.7), we could not guarantee that the corrections and the incertitude of the initial data would not spoil the relation (15.7) and turn the flow equation into an unmanageable relation of little use.

References

[A1]   Anderson, P. *"Luttinger-liquid" behaviour of the normal metallic state of the 2D Hubbard model*, Physical Review Letters **64**, 1839–1841, 1990.

[A2]   Anderson, P. *The "infrared catastrophe": When does it trash Fermi liquid theory?* Preprint #405, February 1993, Princeton, N.J.

[ADG]   Abrikosov, A. A., Gorkov, L. P., and Dzyaloshinski, I. E. *Methods of quantum field theory in Statistical Physics*, Dover Publications, 1963.

[Ai1]   Aizenman, M. *Translation invariance and phase coexistence in the two-dimensional Ising system*, Communications in Mathematical Physics **73**, 83–94, 1980.

[Ai2]   Aizenman, M. *Geometric analysis of ϕ_4^4 fields and Ising model*, Communications in Mathematical Physics **86**, 1–48, 1982.

[B]   Benfatto, G. *Renormalization group approach to zero temperature Bose condensation*, Proceedings, "Workshop on Constructive Results in Field Theory, Statistical Mechanics and Condensed Matter Physics." Palaiseau, July 25-27, 1994 (to be published).

[BD]   Bloch, C., and De Dominicis, C. *Un développement asymptotique du potentiel de Gibbs d'un système d'un grand nombre de particules*, Nuclear Physics **7**, 459–479, 1957.

[BG1]   Benfatto, G., and Gallavotti, G. *Perturbation theory of the Fermi surface in a quantum liquid. A general quasi particle formalism and one-dimensional systems*, Journal of Statistical Physics **59**, 541–664, 1990.

[BG2]   Benfatto, G., and Gallavotti, G. *Renormalization group approach to the theory of the Fermi surface*, Physical Review **B42**, 9967–9972, 1990.

[BGM]   Benfatto, G., Gallavotti, G., and Mastropietro, V. *Renormalization group and the Fermi surface in the Luttinger model*, Physical Review **B45**, 5468–5480, 1992.

[BGPS]   Benfatto, G., Gallavotti, G., Procacci, A., and Scoppola, B. *Beta function and Schwinger functions for many fermion systems in one dimension. Anomaly of the Fermi surface*, Communications in Mathematical Physics **160**, 93–172, 1994.

[BGr]   Benfatto G., and Gruber, C. *On the susceptibility and clustering properties of unbounded spins*, Journal of Statistical Physics **37**, 237–256, 1984.

[BM]   Bonetto, F., and Mastropietro, V. *Renormalization group in a $d = 1$ system of interacting spinning fermions in a periodic potential*, CARR preprint, 1994 (to be published in Communications in Mathematical Physics).

[Bo] Bogoliubov, N. N. Journal of Physics (USSR) **11**, 23–32, 1947.

[BS1] Bleher, P., and Sinai, Y. *Investigation of the critical point in models of the type of Dyson's hierarchical models*, Communications in Mathematical Physics **33**, 23–42, 1973.

[BS2] Bleher, P., and Sinai, Y. *Critical indices for Dyson's asymptotically hierarchical models*, Communications in Mathematical Physics **45**, 247–278, 1975.

[C] Carleson, L. *On the convergence and growth of partial sums of Fourier series*, Acta Mathematica **116**, 135–157, 1966.

[CE] Collet, P., and Eckmann, J. P. *A renormalization group analysis of the hierarchical model in statistical mechanics*, Vol. 74, Lecture Notes in Physics, Springer-Verlag, 1974.

[CW] Coleman, S., and Weinberg, E. *Radiative corrections as the origin of spontaneous symmetry breaking*, Physical Review **D7**, 1888–1910, 1973.

[D] Dyson, F. *An Ising ferromagnet with distribution of long range order*, Communications in Mathematical Physics **21**, 269–283, 1971.

[DaV] Da Veiga, P. *Construction de modèles nonrenormalizables en théorie quantique des champs*, Thèse, Universitè d'Orsay, 1991.

[DCR] De Calan, C., and Rivasseau, V. *Local existence of the Borel transform in euclidean ϕ_4^4*, Communications in Mathematical Physics **82**, 69–100, 1981.

[DJ] Di Castro, C., and Jona Lasinio, G. *On the microscopic foundations of the scaling laws*, Physics Letters **29A**, 322–323, 1969.

[F] Fefferman, C. *Pointwise convergence of Fourier series*, Annals of Mathematics **98**, 551–571, 1973.

[Fe] Felder, G. *Construction of a nontrivial planar field theory with ultraviolet stable fixed point*, Communications in Mathematical Physics **102**, 139–155, 1985.

[FHRW] Feldman, J., Hurd, L., Rosen, T. R., and Wright, J. D. *QED: A Proof of renormalizability*, Lecture Notes in Physics, **312**, Springer-Verlag, 1988.

[FMRT1] Feldman, J., Magnen, J., Rivasseau, V., and Trubowitz, E. *An infinite volume expansion for many fermion Green's functions*, Helvetica Physica Acta **65**, 679–721, 1992.

[FMRT2] Feldman, J., Magnen, J., Rivasseau, V., and Trubowitz, E. *Ward identities and a perturbative analysis of a $U(1)$ Goldstone Boson in a many fermion system*, Helvetica Physica Acta **66**, 498–550, 1993.

[FO] Feldman, J., and Osterwalder, K. *The Wightman axioms and the mass gap for weakly coupled ϕ_3^4 quantum field theories*, Annals of Physics **97**, 80–135, 1976.

[Fr] Fröhlich, J. *On the triviality of $\lambda\phi_d^4$ theories and the approach to the critical point in $d \geq 4$ dimensions*, Nuclear Physics **B200**, 281–296, 1982.

[FS] Fröhlich, J., and Spencer, T. *The Kosterlitz–Thouless transition in two dimension abelian spin systems and the Coulomb gas*, Communications in Mathematical Physics **81**, 527–602, 1981.

[FT1] Feldman, J., and Trubowitz, E. *Perturbation theory for many fermion systems*, Helvetica Physica Acta **63**, 157–260, 1990.

[FT2] Feldman, J., and Trubowitz, E. *The flow of an electron-positron system to the superconducting state*, Helvetica Physica Acta **64**, 213–357, 1991.

[G1] Gallavotti, G. *Renormalization theory and ultraviolet stability via renormalization group methods*, Reviews of Modern Physics **57**, 471–569, 1985.

[G2] Gallavotti, G. *Some aspects of renormalization problems in statistical mechanics*, Memorie dell' Accademia dei Lincei **15**, 23–59, 1978.

[G3] Gallavotti, G. *Quasi particles and scaling properties at the Fermi surface*. In "Probabilistic Methods in Mathematical Physics," ed. F. Guerra, M. Loffredo, and C. Marchioro, Quaderno CNR, 222–232, World Scientific, 1992.

[Gi] Ginibre, J. *On the asymptotic exactness of the Bogoliubov approximation for many boson systems*, Communications in Mathematical Physics **8**, 26–51, 1968.

[GJ] Glimm, J., and Jaffe, A. *Quantum Physics*, Springer–Verlag, 1981.

[GJS] Glimm, J., Jaffe, A., and Spencer, T. *The Wightman axioms and particle structure in the $P(\phi)_2$ quantum field model*, Annals of Mathematics **100**, 585–632, 1974.

[GK1] Gawedzky, K., and Kupiainen, A. *Gross Neveu model through convergent perturbation expansion*, Communications in Mathematical Physics **102**, 1–30, 1985.

[GK2] Gawedzky, K., and Kupiainen, A. *Renormalization of a nonrenormalizable quantum field theory*, Nucear Physics **B262**, 33–48, 1985.

[GK3] Gawedzky, K., and Kupiainen, A. *Massless lattice ϕ_4^4 theory: Rigorous control of a renormalizable asymptotically free model*, Communications in Mathematical Physics **99**, 197–252, 1985.

[GK4] Gawedzky, K., and Kupiainen, A. *Block spin renormalization group for dipole gas and $(\partial\phi)^4$*, Annals of Physics **147**, 198–243, 1983.

[GN1] Gallavotti, G., and Nicolò, F. *Renormalization theory for four dimensional scalar fields, I*, Communications in Mathematical Physics **100**, 545–590, 1985.

[GN2] Gallavotti, G., and Nicolò, F. *Renormalization theory for four dimensional scalar fields, II*, Communications in Mathematical Physics **101**, 1–36, 1985.

[Go] Gosselin, J. *On the convergence of Walsh-Fourier series for $L^2(0,1)$*. In Studies in Analysis, pg. 223-232, Advances in Mathematics, Supplementary Studies **4**, 1979.

[GR] Gallavotti, G., and Rivasseau, V. *φ^4 field theory in dimension four. A modern introduction to its unsolved problems*, Annales de l'Institut

Henri Poincaré **A40**, 185–220, 1984.

[Gr]   Gross, L. *Decay of correlations in classical lattice models at high temperature*, Communications in Mathematical Physics **68**, 9–27, 1979.

[Gu]   Guerra, F. *Uniqueness of the vacuum energy density and Van Hove phenomena in the infinite volume limit for two-dimensional self-coupled Bose fields*, Physical Review Letters **28**, 1213–1215, 1972.

[H]   Hepp, K. *Théorie de la renormalization*, Vol. 2 of Lecture Notes in Physics, Springer-Verlag, 1966.

[Hi]   Higuchi, Y. *On the absence of nontranslation invariant Gibbs states for the two-dimensional Ising model*, In Random Fields: Rigorous Results in Statistical Mechanics and Quantum Field Theory (Colloquium at Ezstergom, June 1979), ed. J. Fritz, J. Lebowitz, and D. Szasz, North Holland, 1982.

[Ho]   't Hooft, G. *Which topological features of a gauge theory can be responsible for permanent confinement?* In Recent Developments in Gauge Theories (Cargèse, 1979), ed. G. 't Hooft et al., Plenum Press, 1980.

[HP]   Hugenholtz, N. M., and Pines, D. *Ground-state energy and excitation spectrum of a system of interacting bosons*, Physical Review **116**, 489, 1959.

[IZ]   Itzykson, C., and Zuber, J.-B. *Quantum Field Theory*, McGraw-Hill, 1980.

[J]   Jona-Lasinio, G. *Relativistic field theories with symmetry-breaking solution*, Nuovo Cimento **34**, 1790–1795, 1964.

[JMMS]   Jimbo, M, Miwa, T., Mori, Y., and Sato, M. *Density matrix of an impenetrable gas and the fifth Painlevé trascendent*, Physica **D1**, 80–158, 1980.

[KLS]   Kennedy, T., Lieb, E., and Shastry, B. S. *The XY model has long-range order for all spins and all dimensions greater than one*, Physical Review Letters **61**, 2582–2584, 1988.

[KW1]   Koch, H., and Wittwer, P. *A nongaussian renormalization group fixed point for hierarchical scalar lattice field theories*, Communications in Mathematical Physics **106**, 495–532, 1986.

[KW2]   Koch, H., and Wittwer, P. *On the renormalization group transformation for scalar hierarchical models*, Communications in Mathematical Physics **138**, 537, 1991.

[L]   Luttinger, J. *An exactly soluble model of a many fermion system*, Journal of Mathematical Physics 4, 1154–1162, 1963.

[LL]   Lieb, E., and Lininger, W. *Exact analysis of an interacting Bose gas. I. The general solution and the ground state*, Physical Review **130**, 1605–1624, 1963.

[LM]   Lieb, E., and Mattis, D.*Mathematical Physics in One Dimension*, Academic Press, New York, 1966.

[LW]   Luttinger, J., and Ward, J. *Ground state energy of a many fermion system*, Physical Review **118**, 1417–1427,1960.

[McK]   MacKay, R. *A renormalization approach to invariant circles in area preserving maps*, Physica **D7**, 283–300, 1983.

[ML]   Mattis, D., and Lieb, E. *Exact solution of a many fermions system and its associated boson field*, Journal of Mathematical Physics **6**, 304–312, 1965. Reprinted in [LM].

[MS]   Magnen, J., and Sénéor, R. *The infinite volume limit of the ϕ_3^4 model*, Annales de l'Institut Henri Poincaré **A24**, 95–159, 1976.

[N]   Nelson, E. *Construction of quantum fields from Markoff fields*, Journal of Functional Analysis **12**, 97–112, 1973.

[NN]   Nepomnyashchii, Yu. A., and Nepomnyashchii, A. A. *Infrared divergence in field theory of a Bose system with a condensate*, Soviet Phys. JETP **48**, 493-501, 1978.

[P]   Parisi, G. *Statistical Field Theory*, Addison-Wesley, 1987.

[Po1]   Polchinski, J. *Effective field theory and the Fermi surface*, University of Texas, preprint UTTC-20-92.

[Po2]   Polchinski, J. *Renormalization group and effective lagrangians*, Nuclear Physics **B231**, 269–295, 1984.

[PS]   Popov, V. N., and Seredniakov, A. V. *Low-frequency asymptotic form of the self-energy parts of a superfluid Bose systemat $T = 0$*, Soviet Phys. JETP **50**, 193–195, 1979.

[R]   Rivasseau, V. *From perturbative to constructive renormalization*, Princeton Series in Physics, Princeton Univerity Press, 1991.

[Rm1]   Benfatto, G., Cassandro, M., Gallavotti, G., Nicolò, F., Presutti, E., Olivieri, E., and Scacciatelli, E. *Some probabilistic techniques in field theory*, Communications in Mathematical Physics **59**, 143–166, 1978.

[Rm2]   Benfatto, G., Cassandro, M., Gallavotti, G., Nicolò, F., Presutti, E., Olivieri, E., and Scacciatelli, E. *Ultraviolet stability in euclidean scalar field theories*, Communications in Mathematical Physics **71**, 95–130, 1980.

[Ru]   Russo, L. *The infinite cluster method in the two-dimensional Ising model*, Communications in Mathematical Physics **67** 251–266, 1979.

[Sh]   Shankar, R. *Renormalization group approach to interacting fermions*, Preprint, Sloan Lab., Yale University, 1992.

[SK]   Shenker, S. J., and Kadanoff, L. P. *Critical behaviour of a KAM surface: I, Empirical results*, Journal of Statistical Physics **27**, 631–656, 1982.

[So]   Sólyom J. *The Fermi gas model of one-dimensional conductors*, Advances in Physics **28**, 201–303, 1979.

[T]   Tomonaga, S. *Remarks on Bloch's method of sound waves applied to many fermion problems*, Progress of Theoretical Physics 5, 544–569, 1950. Reprinted in [LM].

[VT]   Vaidya, H. G., and Tracy, C. A. *One-particle reduced density matrix of impenetrable bosons in one-dimension at zero temperature*, Journal of Mathematical Physics **20**, 2291–2312, 1979.

[W1] Wilson, K. G. *Model of coupling constant renormalization*, Physical Review **D2**, 1438–1472, 1970.

[W2] Wilson, K. G. *Renormalization group and strong interactions*, Physical Review **D3**, 1818–1846, 1971.

[W3] Wilson, K. G. *Renormalization of a scalar field in strong coupling*, Physical Review **D6**, 419–426, 1972.

[W4] Wilson, K. G. *Quantum field theory models in less than four dimensions*, Physical Review **D7**, 2911–2926, 1973.

[W5] Wilson, K. G. *The renormalization group and critical phenomena*, Nobel Prize lecture, Reviews of Modern Physics **55**, 583–600, 1983.

[W6] Wilson, K. G. *The renormalization group: Critical phenomena and the Kondo problem*, Reviews of Modern Physics **47**, 773–840, 1975.

[We] Weinberg, S. *Effective action and renormalization group flow of anisotropic superconductors*, University of Texas preprint, UTTG-18-93.

[WF] Wilson, K. G., and Fisher, M. *Critical exponents in 3.99 dimensions*, Physical Review Letters **28**, 240–243, 1972.

[WG] Wightman, A., and Gärding, L. *Fields as operator-valued distributions in relativistic quantum theory*, Arkiv för Fysik **28**, 129–189, 1965.

[WK] Wilson, K. G., and Kogut, J. B. *The renormalization group and the ε-expansion*, Physics Reports **12**, 76–199, 1974.

Subject Index

Citation Index